ANALYTICAL PHILOSOPHY OF TECHNOLOGY

BOSTON STUDIES IN THE PHILOSOPHY OF SCIENCE

EDITED BY ROBERT S. COHEN AND MARX W. WARTOFSKY

VOLUME 63

FRIEDRICH RAPP

*Institut für Philosophie, Wissenschaftstheorie, Wissenschafts- und
Technikgeschichte, Technische Universität Berlin*

ANALYTICAL PHILOSOPHY OF TECHNOLOGY

Translated by Stanley R. Carpenter
and
Theodor Langenbruch

D. REIDEL PUBLISHING COMPANY

DORDRECHT : HOLLAND / BOSTON : U.S.A.
LONDON : ENGLAND

Library of Congress Cataloging in Publication Data

Rapp, Friedrich.
Analytical philosophy of technology.

(Boston studies in the philosophy of science ; v. 63)
Translation of: Analytische Technikphilosophie.
Includes bibliographical references and index.
1. Technology–Philosophy. I. Title. II. Series.
Q174.B67 vol. 63 [T14]   501s   [601]   81-201
ISBN 90-277-1221-2   AACR2
ISBN 90-277-1222-0 (pbk.)

Published by D. Reidel Publishing Company,
P.O. Box 17, 3300 AA Dordrecht, Holland.

Sold and distributed in the U.S.A. and Canada
by Kluwer Boston Inc.,
190 Old Derby Street, Hingham, MA 02043, U.S.A.

In all other countries, sold and distributed
by Kluwer Academic Publishers Group,
P.O. Box 322, 3300 AH Dordrecht, Holland.

D. Reidel Publishing Company is a member of the Kluwer Group.

Originally published in German by
Karl Alber Verlag

All Rights Reserved
This translation copyright © 1981 by D. Reidel Publishing Company, Dordrecht, Holland
No part of the material protected by this copyright notice may be reproduced or
utilized in any form or by any means, electronic or mechanical,
including photocopying, recording or by any informational storage and
retrieval system, without written permission from the copyright owner

Printed in The Netherlands

EDITORIAL PREFACE

Friedrich Rapp, in this magisterial and critical essay on technology, the complex human phenomenon that demands philosophy of science, philosophy of culture, moral insight, and historical sensitivity for its understanding, writes modestly of the grave and tentative situation in the philosophy of technology. Despite the profound thinkers who have devoted time and imagination and rational penetration, despite the massive literature now available, despite the varied and comparative viewpoints of political, analytic, metaphysical, cultural, even esthetic commitments, indeed despite the honest joining of historical and systematic methods of investigation, we are far from a satisfactory understanding of the joys and sorrows, the achievements and disappointments, of the technological saga of human societies. Professor Rapp has prepared this report on the philosophical understanding of technology for a troubled world; if ever philosophy were needed, it is in the practical attempt to find alternatives among technologies, to foresee dangers and opportunities, to choose with a sense of the possibility of fulfilling humane values. Emerson spoke of the scholar not as a specialist apart, but as 'Man thinking' and Rapp's essay so speaks to all of us, industrial world or third world, engineers or humanists, tired or energetic, fearful or optimistic.

We understand from Rapp, and from many others, that the problems posed here are really not due to technology: they may be emphasized, transformed, distorted, perhaps transcended, perhaps degraded, by what scientists and engineers and entrepreneurial managers have wrought, but they are ancient problems of dominations, values, decisions, and struggles. They certainly need, and will continue to require, conceptual clarification, that sorting out and drawing of distinctions, which a philosopher may contribute to the theoretical and practical labors of technical workers and to our intellectual equipment for public policy. We hope this book will help, both by the author's lucid analytic survey of so much of the classical and recent research on technology and society, and by his own critical contributions.

* * *

We thank Professors Stanley Carpenter and Theodor Langenbruch for their devoted work on the English translation. Stanley Carpenter himself has contributed serious papers to the literature on philosophy of technology. We also are grateful to Renate Hanauer for her research and editorial work on this book.

March 1981   ROBERT S. COHEN
Boston University Center   MARX W. WARTOFSKY
for the Philosophy and
History of Science

TABLE OF CONTENTS

EDITORIAL PREFACE   v

PREFACE TO THE ENGLISH EDITION   xi

I. *Stages of the Philosophy of Technology*   1
   1. Introduction   1
   2. The Engineering Perspective   4
   3. Philosophy of Culture   8
   4. Social Criticism   13
   5. The Earth as a System   16
   6. The Problem of Diversity of Approaches   19

II. *Differing Versions of the Concept of 'Technology'*   23
   1. Problems of Definition   23
   2. Historical and Systematic Analysis   24
   3. Periods in the History of Technology   26
   4. Semantic Variations of the Concept of 'Technology'   31
   5. Attempts at Definition   33

III. *Methodological Analysis*   37
   1. The Determinants of Technological Development   37
   2. The Range of Action   42
   3. The Transformation of the Material World   47
   4. The Neutrality of Technological Means   53
   5. Hypothetical Imperatives   57
   6. Technological Progress   62

IV. The Road to Modern Technology — 66
1. The Socio-cultural Approach — 66
2. Historical Determination — 67
3. Magical and Technological Thinking — 70
4. Socio-economic Conditions — 74
5. Technological Foundations — 80
6. The Industrial Revolution — 84
7. Engineering Sciences and Natural Sciences — 87
8. Intellectual Prerequisites — 92
   a. Valuation of Work — 94
   b. Efficient Management — 96
   c. The Impulse for Technological Creativity — 97
   d. Rational Thought and the Enlightenment — 99
   e. Objectification of Nature — 101
   f. The Mechanistic View of Nature — 103
   g. The Mathematical Model — 105
   h. Experimental Investigations — 106
9. Complex Interconnections — 108
10. Natural Instinct and Volitional Creativity — 110

V. The Technological World — 118
1. Nature and Artifacts — 118
2. The Cosmic Dimension — 125
3. Accumulation and 'Self-Reinforcement' — 133
4. The Acting Individuals — 140
5. Individual Freedom and Collective Tasks — 143
6. The Universality of Modern Technology — 152
   a. The Transformation of the Physical World — 152
   b. The Changed Situation of Life — 154
   c. Worldwide Expansion — 160
7. The Benefits of Technology and Their Cost — 165
8. Changed Criteria — 171
9. New Values — 176
10. The Crisis in the Assessment of Technology — 180

BIBLIOGRAPHY	187
NAME INDEX	195
SUBJECT INDEX	197

PREFACE TO THE ENGLISH EDITION

The translation of my book *Analytische Technikphilosophie* into English constituted an opportune occasion for revising the text. Alterations have been restricted to occasional rephrasing in order to make more clear and more explicit the meaning of some formulations, and to condense passages where desirable. Neither the reviews of the German edition published hitherto nor the recent development of the field have called for major changes.

In the book, new approaches, starting from a problem-centered systematic survey of the actual state of the art and a critical discussion of the solutions available up to now, are put forward. The summary of the historical development given in the first chapter and the majority of the titles quoted refer to (West) German research. But this does not restrict the scope of investigation. Major points of the international discussion are explicitly taken into account. As far as specific questions are concerned, the German investigations can be read as exemplifying the general problems. Thus, the four historical stages treated in chapter I are in one way or another present in any discussion of the philosophical problems of technology.

The German philosophy of technology, often regarded as pioneering in the field, has been associated with the metaphysical interpretations of Dessauer and Heidegger. Approaches of this type are also considered here, but they do not constitute the only topic of the book. It seems that with respect to the philosophical problems of technology a situation similar to other fields of philosophical investigation (philosophy of science, philosophy of history) obtains: modern science-based technology and the features of the world shaped by it are so complex that they cannot adequately be dealt with by means of *a priori* considerations which neglect the *a posteriori* facts. It is only after having analyzed the philosophically relevant traits of the historical development and the empirically given systematic features of technology that a well-founded metaphysical interpretation becomes possible. In accordance with this line of reasoning the first chapters of the book prepare the ground for the metaphysical explanation of technology discussed at the end of chapter IV and in chapter V.

In spite of the increasing progress being made in the field*, a well-elaborated philosophy of technology – let alone a generally accepted paradigm – is still more a desideratum than a concrete reality. The major obstacle to a simple treatment is the 'totality' of modern technology, as stressed, for example, by Ellul and also by Marcuse. Indeed, today there is hardly an area of individual and social life which is not influenced by technological change. The task of philosophy consists not only in drawing attention to this circumstance but in making explicit what precisely this totality consists in, how it could have come about, and what consequences follow from it. Only by going into the details will it be possible to arrive at fruitful hypotheses, the closer inspection and critical discussion of which will help to clarify the situation and to suggest solutions. It is in terms of this analytical approach that, for example, in sections III.2., IV.8., and V.3., the elements relevant for the range of action, for the historical development, and for the dynamic character of modern technology are discussed. In this way the relationships obtaining between these elements and their influence on the 'mechanism' of the whole process may become discernible.

---

* For detailed information see the yearbooks *Research in Philosophy and Technology*, ed. by P. Durbin, starting with vol. 1, Greenwhich, Conn., 1978, and my forthcoming survey 'Philosophy of Technology' in *Contemporary Philosophy*, vol. 2., ed. by F. Floistad, The Hague, 1981.

PREFACE TO THE ENGLISH EDITION

In addition to considering the empirical data and in addition to the analytical method applied, putting an accent on the engineering sciences constitutes a further characteristic trait of the enquiry put forward here. This option seems legitimate, if one considers the fact that the concrete material artifacts, and the procedures by means of which they are brought about and put to use, make up the kernel of modern technology. Without reference to the engineering approach and the processes of research and development derived from it the dynamics of modern technology can hardly be explained. This does not imply that cultural, social, or economic factors are irrelevant. But their concrete influence is always shaped by the state of engineering. The process of accumulation and of 'self-reinforcement' of modern technology as analyzed in section V.3. is also based on the structure of the engineering sciences.

In works on the problems of modern technology addressed to a broader public, the author's (optimistic or pessimistic) opinion about the prospects of future development often comes into play. No such thing is intended here. As far as philosophical analysis is concerned, the personal commitment must be disciplined by and translated into objectifying scientific categories. To rely on an intuitive act of decision to solve a philosophical dispute about the 'essence' of modern technology, or about the normative ethics we need for managing the future of mankind, would amount to giving up conceptual understanding and rational argument. What we need is a scrutinized analysis of the situation obtaining. Once this has been achieved, one might easily arrive at a well-founded evaluation in terms of the theory of decision by ascribing subjective values to the relevant factors. Clearly this is only a stylized scheme. But it can help to exhibit the complementary character of 'subjective' evaluation and the 'objective' state of affairs. The relation obtaining here may perhaps best be expressed by adapting a phrase of Kant's: Knowledge without commitment is empty, commitment without knowledge is blind.

I would like to thank my friends Stan Carpenter and Theo Langenbruch for having so perfectly achieved the intricate and troublesome task of translating the features of the German version into clear and readable English (the result being that sometimes I found my ideas expressed more precisely by the translation than in my German wording). My thanks go as well to Professor Robert S. Cohen and Ms. Renate Hanauer of Boston University and to Mrs. J. C. Kuipers of the D. Reidel Publishing Company. It is due to their constructive collaboration that this edition appears.

FRIEDRICH RAPP

CHAPTER I

STAGES OF THE PHILOSOPHY OF TECHNOLOGY

1. INTRODUCTION

While human history has been marked by a great variety of advanced cultures, the material conditions of life within each one of them have remained unchanged for long stretches of time. Although these cultures developed the arts and crafts to a high degree of perfection, man's action upon nature was limited to relatively simple techniques based on traditional skills and elementary knowledge. With the advent of the Industrial Revolution, however, begun in England about two hundred years ago, a fundamental change occurred. The invention of new sources of mechanical energy, the use of advanced construction materials, and the rationalization of production processes combined to initiate developments which have continued to the present. In the nineteenth century, traditional technologies based on artisan skills were fused with the physical sciences which had developed from natural philosophy. The result was an all-encompassing scientific and technological process which, by virtue of its inherent dynamism, has gained momentum ever since. There is every indication that mankind stands at the threshold of a technological age possessing not only undreamed-of material possibilities but also cause for anxiety.

Today, directly or indirectly, technology influences all areas of individual and social life. In the industrialized countries, living standards and improved modes of transportation and communication are based as much on technology, as are increases in life expectancy and the rapid rate of social change. While there is a wide-spread tendency to accept unreflectively the accomplishments and conveniences of technology, its negative consequences are more likely to lead to reflection and criticism. There are, for example, many complaints about the division of labor, depersonalization of the work process, and the drive for efficiency which have led, both in the realm of work and of interpersonal relationships, to man's alienation from an existence which is readily intelligible and inwardly fulfilling. Yet it appears we are fated to live with technology which, while always the result of human actions, has gained a momentum of its own and seems to develop almost uncontrollably. Progress in medical technology in the developing countries, for example, has led to a population explosion with accompanying food shortages which can only be coped with through the introduction of such technological aids as birth control devices, artificial fertilizers, and agricultural machinery. Mankind as a whole faces the threat of nuclear annihilation because of the continuing arms race. At the same time, the destruction of the natural environment and crises of both natural resources and energy make it clear that technological development cannot proceed without limits.

To a large extent, the dynamics of change have left theoretical understanding behind. This is why technology, although intentionally and systematically created by man, is now experienced as an alien and uncanny force. There are two reasons for this unsatisfactory situation. In the first place, the structure of technology is itself *multi-dimensional* and *complex*. Its manifold varied aspects can no more be summarized by concise formulae than can, say, 'science' or 'politics.' Paradoxically enough, the other reason is related to the specific *theoretical* bias of Western thought itself. But demystifying the world, rationalizing economic processes, and developing the methodology of the exact sciences, this style of thinking made modern technology possible. From its origins in Europe, it has spread to the entire world. Yet throughout this process, the traditional preference for detached theoretical reflection over practical action has remained. Because of this, technology has been devalued as an object for critical scrutiny. Thus, although its origins reflect the spirit of enlightened scientific rationality — and in a broader sense the Western philosophical tradition itself — and although its efficacy is obvious, modern technology has only belatedly become the object of basic philosophical investigations. In fact, with the exception of Marxist philosophy, the tendency to cling to the traditional definition of man as *animal rationale* continues to lead philosophy to neglect the concept of man as *homo faber*, an aspect of crucial importance today.

It would be an error, however, to blame only the traditional humanistic model of education for the lack of sufficient reflection upon the problems of modern technology. Indeed in reaction to this model, there is generally a stronger emphasis on the scientific-technical education. This has meant, more often than not, that a contemplative humanism is replaced by a purely success-oriented technicism which in turn has contributed to the unfortunate confrontation between the sciences and humanities as 'two cultures'. Inappropriate exaggerations aside, however, it is fair to say that these two points of view are by no means mutually exclusive. To the contrary, we are dealing here with a necessarily complementary relationship. When discussing the *meaning* of technological actions or the *criteria* of technical decision-making, we cannot avoid a 'humanistic' reflection on values. If, on the other hand, the issue is one of determining the range of technically possible *solutions* for a given problem or of predicting the physical *consequences* of any particular technological decision, it is only the scientist or engineer who can competently inform us.

In the civilizations preceding the Industrial Revolution, technology played a subordinate role in life and culture. Accordingly, there was no obvious need for separate philosophical treatment of matters pertaining to it. It is not until the end of the nineteenth century that one finds, in the German language, the beginnings of a distinct and independent philosophy of technology. While it is not possible to summarize the developments which followed in terms of a single scheme (cf. Huning *1.16*; Lenk-Ropohl *1.20*; and Moser *1.23*), we may, by way of a broad overview, identify four approaches. For a time, each dominated philosophical discussion, though not to the exclusion of other viewpoints. These four vantage points are engineering science, cultural philosophy, social criticism, and systems theory, and their succession roughly parallels the transition from the technological optimism of the nineteenth century to the more critical attitude of present times. Because each of the four thematic areas focuses on one particular aspect of technology, the discussion of them also sheds some light on systemic questions pertaining to technology.

2. THE ENGINEERING PERSPECTIVE

The outward signs of technology, instruments, machines, and other devices, stand out in sharp relief from objects of the inorganic and organic world. All such artifacts owe their existence to having first been thought out by man and then systematically and suitably made. Thus it was only natural that philosophical reflection first focused on two indispensable prerequisites to the emerging machine technology, namely, the creative act of invention and the role of the engineer.

The geographer and philosopher Ernst Kapp is considered the founder of the philosophy of technology. In his *Grundlinien einer Philosophie der Technik* (*1.19*), published in 1877, Kapp explains technical inventions as the material embodiments of the imagination, and technological activity as "organ projection." The hand is for him the model for all artifacts, the archetypal tool. So, for example, he interprets the hammer as an imitation of the arm with a clenched fist. Like most nineteenth century authors, Kapp is decidedly optimistic about the potential of technology, viewing it as means of cultural, moral, and intellectual advancement and "self-redemption" (*1.19*, v). Kapp's theses of organ projection and self-redemption through technology have since been modified and extended by Arnold Gehlen (*5.9*, 93–95)* and Donald Brinkmann (*1.8*, 105–106).

For the engineer Max Eyth, the essence of technological action is the creation of new material products. In his broadly conceived analysis, he first divides the inventive process into three phases: (1) Germination of the idea through a process of intuitive inspiration. Here, the inventor acts without outside stimulus. He is driven by the same creative urge and joy of creation which inspire the artist. (2) Materialization. The idea is made concrete in a prototype which tests the feasibility of the design concept. A favorable outcome leads to gradual refinement of the prototype. This laborious process may even call for further inventions until the product is finally at the point of practical utility. (3) The last phase consists in the successful dissemination and utilization of the invention (*3.4*, 262–267).

In 1906, Eyth's ideas were systematized and further elaborated by the patent lawyer and electrical engineer Alard du Bois-Reymond in *Erfindung und Erfinder* (*3.2*). He distinguishes between the process of invention and the resulting product, sharply separating the act of invention, a psychological event, from its objective embodiment. While the objective content of possible combinations of material elements is in principle timeless, it requires the act of invention to be brought to human awareness. Potentially, for

---

* The number in italics refers to the title listed in the Bibliography at the end of this book. The figures after the comma refer to the page numbers. The volume number, where a work consists of more than one volume, is indicated by a Roman numeral.

example, the production of gun powder always existed. The Chinese, as a matter of fact, were using it long before the Europeans. By the same token, all inventions yet to be made already exist as possible material configurations, although we are not presently aware of them. The creative intellectual achievement consists exclusively in the act of recognition. Considered by itself as a material object, the recognized invention is nothing but a brute combination of material elements. Every invention consists in pointing out certain technological possibilities that agree with particular human needs. Consequently, we are dealing with an invention whenever a heretofore unknown coincidence is discovered between material properties and human needs.

The idea that technological possibilities exist in latent form is speculatively expanded by Friedrich Dessauer in his 1927 *Philosophie der Technik* (*1.10*). Dessauer, himself an engineer and X-ray specialist, starts with the assumption that for every uniquely-defined technological problem there exists, for every given time and setting, one and only one optimal solution. This solution, then, is found, not created. Any improvement of an actual invention asymptotically approaches this pre-established theoretical ideal. Dessauer combines this Platonic concept of pre-existing perfect solutions with a theological interpretation. By actualizing the potential being of pre-existent technological forms, man carries on creation, as it were. God avails himself of technological man to continue the Creation. This view is maintained by Dessauer in his later comprehensive work *Streit um die Technik* (*1.11*). (Cf. the detailed assessment in Klaus Tuchel, (*1.30*).)

Dessauer concedes that the increasing complexity of technological products and processes has led to a progressive reduction in the importance of "pioneering" inventions based on the creative achievements of solitary individuals. Their place has been taken by "developmental" inventions which, because they reflect the existing state of technological knowledge and know-how, frequently originate independently at different places. Such inventions, which are caused both by the scientific and technological 'spirit of the times' and by existing human needs, are usually created by a team of experts from different fields who follow a detailed work plan to progressively refine the product as desired (*1.11*, 168–169).

Developmental inventions demonstrate that modern technology cannot be reduced to either inventive skills alone or to mere application of scientific knowledge. Modern technology results from joint efforts of engineers (and scientists, as required by circumstances) who apply their professional knowledge and expertise to the creation and gradual development of a product up to the point of routine production. This trend toward division of labor and systematization has consequently led to the establishment of appropriate disciplines within the engineering sciences. Cybernetics and systems engineering, for example, deal with subject matters and procedures utilized in the analysis and development of modern technological systems.

It is noteworthy that there have recently been detailed studies of the methodologies of the sciences, a field closely related to technology. Investigations in the analytical philosophy of science have contributed significant insights concerning, for example, the structure of scientific theories and the different types of scientific explanation. While the field of technology is not as clearly demarcated as that of science, one increasingly finds comparable issues discussed, reflecting a keener awareness of methodological questions. With technology, however, things are not as clear-cut as with the sciences. In analyzing science, it is possible, without too much oversimplification, to concentrate on the logical analysis and factual content of systems of statements in abstraction from concrete acts and processes as well as from the social context. An analogous approach to technology would eliminate all those aspects which go beyond the methodological analysis of goal-oriented activity. This would result in a great simplification of the actual situation since, in addition to its impact on technical solutions themselves, modern technology also profoundly affects the concrete every-day world, social change, and the biological environment. Nevertheless, such analyses are indispensable if we are to grasp the methodological status — and beyond that the epistemological status — of technological actions. Tadeusz Kotarbiński (*3.6*) has given a general analysis of the structure and various forms of effective practical action. (Relevant literature is compiled in Friedrich Rapp (*3.7*).)

3. PHILOSOPHY OF CULTURE

Only recently did technology become a topic of research for the philosophy of culture. The writings of Friedrich Nietzsche and Wilhelm Dilthey as well as Eduard Spranger and Ernst Rothacker, reflecting a humanistic or historical orientation, contain no detailed discussion of technology. (Oswald Spengler's *Man and Technics: A Contribution to a Philosophy of Life* [*Der Mensch und die Technik*] (*1.28*) is much more an essay in the 'philosophy of life' than a systematic treatment of technology.) At this point, one finds no thematic focus on such questions as the relationship between the history of ideas and the development of modern technology, or the role played by technology in man's present self-image, or its importance for his future cultural development.

One of the first significant investigations with such a focus is provided by Max Scheler (*4.26*). He attempts to point out the events in cultural and sociopolitical history which have proved decisive in the origin and development of modern science and technology. Scheler's goal is a synthesis of the interpretations of stark economic materialism on the one hand, and pure cultural idealism on the other. For him, there does not exist unilateral dependency of the practical life on a theoretical understanding, as is the case with Hegel, nor does he accept the converse, as is held by Marx. In Scheler's view, every act of rational cognition is preceded by an emotive act of evaluation, since it is only on the basis of such value-oriented attention that an object appears as significant and worthy of investigation. Accordingly, both the material and ideological components of scientific and technological developments are themselves fundamentally the result of underlying emotive attitudes and predispositions. The urge for power and domination of nature, rather than the concern for utility, is the primary characteristic of modern science and technology. Scheler considers knowledge which leads to and flows from scientific and technological domination to be one of the three basic forms of knowledge, the other two being knowledge acquired through humanistic education, and knowledge of religious salvation. This idea was later revived by Jürgen Habermas in his book *Technik und Wissenschaft als 'Ideologie'* (5.12).

In *Man in the Modern Age* [*Die geistige Situation der Zeit*] (5.18), Karl Jaspers presents a critical interpretation of technology. While rationalization and mechanization in the work sphere and elsewhere have today become indispensable, Jaspers holds that in combination with the mass society they have produced in the average person a matter-of-fact attitude toward things and people with an accompanying loss of personality and individuality. The overpowering authority of technological civilization has transformed personally meaningful existence into a life of mediocrity, dullness, and social role playing. Technology, which tends to separate man from his history and from nature, must therefore not be allowed to become the determining factor in human life. Its proper role should be as an unobtrusive, matter-of-course intermediary. To counterbalance technology, one should "sharpen his awareness of the non-technological to the point of irrefutable clarity" (167).

This approach, which critically juxtaposes technology with the ideal of the intellectually autonomous individual, recurs repeatedly in philosophical literature in a large number of variations. A somewhat contrary position is represented by Ernst Jünger's *Der Arbeiter* (*1.17*). By means of the "worker," the prototype of technological man, technology mobilizes the world. Jünger holds that, while technology undermines all traditional institutions and values, the result is not necessarily deleterious. Nor is technology completely neutral, mere means to any arbitrarily chosen end. Only under the rule of the worker can the dynamic potential of technology become a creative force for development. In contrast to this view, his brother Friedrich Georg Jünger has more recently blamed technology for the irresponsible exploitation of natural resources and destruction of the human spirit, and pointed out the apocalyptic risks of continued technological development (*The Failure of Technology: Perfection without Purpose* [*Die Perfektion der Technik*] (*1.18*).)

In José Ortega y Gasset's *Betrachtungen über die Technik* (*1.25*), the philosophy of technology is placed in the broad context of cultural anthropology. Unlike Dessauer, he does not regard particular inventions as the characteristic features of technology. These, he points out, are always part and parcel of the technological inventory of any particular period. What is decisive is what he calles the "life plan." Man is only biologically determined, not culturally so. Within his particular historical context, he always designs a life plan the ends of which technology serves. In this connection, survival needs are subordinate. What is crucially important is that which can lift man above his biological existence, to a dimension that is "objectively superfluous" (31,64). This technology can do. It is thus inseparably connected with man's capacity for self-creation. To be human is to have a technology. Today's technology, benefiting from the analytical method of modern science, has been perfected to a degree heretofore unimagined. It has thus become a readily available means to whatever goal we choose. All we lack is the creative power to envisage desirable goals for the given technological potential.

The philosophy of technology of Martin Heidegger, as contained in his article *The Question Concerning Technology* [*Die Frage nach der Technik*] (*1.14*), is a special case. In condensed form using esoteric language, Heidegger here presents an interpretation of technology from his standpoint of the metaphysics of being. (To do justice to this viewpoint, one should be famaliar with the Heideggerian corpus.) According to him, we remain blind to the nature of technology both when we either passionaltely accept or reject it, but even more so when we regard it as a mere neutral means to some end. Modern technology challenges nature. This stance contrasts with the preserving, protecting attitude of earlier times. For modern man, the forces of nature constitute merely an arbitrary, disposable resource, or "standing reserve" (*Bestand*). Through technological pro*duction* (her*stellen*), it is in*ducted* (ge*stellt*) into man's service. Yet, Heidegger notes, this instrumental or anthropological interpretation of technology must be complemented by the crucial realization that man is destined by his times to be challenged or appointed (con*ducted*, be*stellt*) to such behavior. In sum, modern technology amounts to a pro*ductive* framework, or "enframing" (*Gestell*). In producing ("revealing") technological objects, man is always in danger of being drawn into the vortex of technology. With this risk, however, man is also afforded the chance to realize the actual essence, the "truth," and to take it into his care.

Arnold Gehlen's anthropological analysis of technology is clearly different from Heidegger's speculative hermeneutics. In his collection of essays titled *Man in the Age of Technology* (*5.8*), dealing with the social-psychological problems of an industrial society, he attacks as academic prejudice the idea that technology is exclusively rationalistic and utilitarian. It is rather for him a necessary consequence of man's inadequate biological endowment. Appropriate tools compensate for deficiencies in human organs. Technical work routines are themselves a kind of closed-loop process controlled by success and automatic in character. By means of engrained habit, man is relieved of the need to master each changing situation anew. Thus, from Gehlen's point of view, technology proves to be primarily a biological and instinctual phenomenon rather than a theoretical or cultural achievement (1–20).

Hannah Arendt arrives at a decidedly negative assessment of the results of modern technology in *The Human Condition* (*4.1*). Antiquity held to the contemplative ideal of life, informed, as it was, by the immediately given, visible world. The present, she notes by contrast, is characterized by a "radical loss of world and reality" as manifested in forsaken and derooted mass society and in the abstract and non-sensuous world view of modern science and even more so in a blind technological hyperactivity which considers everything on the basis of utility and conceives of every action in terms of means and ends only (262–264). When *homo faber* detached himself from the rhythm of organic life and from the flow of history, he lost all certain guideposts and began instead a process of open-ended, indefinite technological activity, in which means undergo continued improvement. In turn, the ends they are intended to serve, whatever they may be, lie completely outside the technological context (305–307).

While Arendt's position is framed on the basis of individualism and Western development, the Brazilian Darcio Ribeiro views technology from the broad perspective of universal history (*4.24*). For him, too, science-based technology is the key to understanding the present situation. Indeed, he sees technology as the "causal factor," the "motor of history." Viewed in the long run, Ribeiro maintains, one may recognize this technological determinism. From a middle range perspective, social institutions are decisive.

In the short run, however, religious, philosophical, and political values ("ideology") are the determining factors. While Ribeiro admits that technology has conjured up the danger of atomic war and led to a general levelling and manipulation of mankind, he sees man's flexibility and adaptability as limitless. Consequently, he believes it will be possible in a future (socialist) order, one in which economic, social, and political life is "rationally" controlled, to make university education the norm, eliminate alienation, and make possible lives of freedom and affluence (196–206). This utopian and optimistic vision of the future leads us to the theme of 'Technology and Society'.

4. SOCIAL CRITICISM

The present discussion focuses in particular on the social problems connected with expanding 'technologization' where, as a rule, social *criticism* predominates. A borderline position between philosophy of culture and social philosophy is represented by Hans Freyer's essay *Über das Dominantwerden technischer Kategorien in der Lebenswelt der industriellen Gesellschaft* (5.6). Today, he notes, technology has become the main frame of reference in the industrialized world. It occupies the place formerly held by religion, art, and politics. Technological expressions such as 'shifting', 'cranking up', 'idling' and 'sand in the works' have become a part of everyday speech. The corresponding evaluative notion of functional efficiency has become the generally accepted norm. All economic and social processes are either directly or indirectly related to technological development. This is as true for concentration of the population in industrial areas, growth of the service sector, and the levelling of consumer tastes, as it is for the formation of interpersonal relations and for social structure and rationalized, purposeful organization. Technological progress, organized and systematized, has become an established institution, subject to its own immanent laws. It functions anonymously, largely independent of accidental conditions and individual achievements as well as of the happiness or suffering of the individual.

Based on the wide-ranging analysis of Jaques Ellul (*5.4*), Helmut Schelsky develops, in *Der Mensch in der wissenschaftlichen Zivilisation* (*5.34*), the idea of the "technological state" with all its ramifications. The starting point of his reflections is the observation that man's present-day relationship to his environment is determined by a ubiquitous technology, while it is man who has produced this artificial technological world, its problems can only be solved by still more technology, systematically planned and appropriately executed. The inner logic of technology itself necessarily subjugates man who thus forfeits his chance to autonomously set goals transcending the technological framework. Scientific and technological imperatives have replaced political action. State and technology thereby combine to form the technological state, the only goal of which consists in the perfecting of means. The result is a "technocracy," yet one in which experts and technologists are not in control. Contrary to expectations, it is technological data alone which dominate and to which expert and politician alike are subject. Schelsky's points have obviously touched on a central problem of modern technology. It is therefore not surprising that the ensuing discussion, first begun in the journal *Atomzeitalter*, continues to this day (cf. Hans Lenk, *5.25*).

A thematically related discussion of modern technology, but with differences in emphasis, exists within the framework of German sociology in the Critical Theory of the 'Frankfurt School' (Theodor W. Adorno, Jürgen Habermas, Max Horkheimer, Alfred Schmidt). Here emancipatory and socio-political involvement and the criticism of existing conditions are emphasized over a mere diagnosis and inventory of organizational structures. The theoretical point of reference in this context, besides the philosophical tradition of Rationalism and the Enlightenment, is mainly

Marx and Hegel. Marx's theory of alienation of the *Economic and Philosophical Manuscripts* (*4.17*b, 63–193) and his socio-economic investigation in *Capital* (*4.17*a) usually serve as the starting point. Horkheimer's work *Eclipse of Reason*, first published in English, 1947, and twenty years later translated into German as *Zur Kritik der instrumentellen Vernunft* (*5.16*), is of particular significance for the philosophy of technology. In this work, he describes the process whereby scientific and technological instruments are increasingly perfected, while, at the same time, objectively binding and rationally founded goals are lost.

Herbert Marcuse, in *One-Dimensional Man* (*5.27*), was to supply a fundamental critique of the general consciousness in the advanced industrial societies. Mercilessly, he points out the evils of resource wastefulness, social exploitation, and the apocalyptic perfecting of armaments. In his opinion, a society must be labelled irrational if there exists, despite an immanent technological efficiency, a glaring incongruity between the concrete realities of life and technological possibilities. Technological progress has become a universally accepted norm. In combination with a sacrosanct demand for productivity and continued growth, it has led, Marcuse holds, to a fusion of culture, politics and economic into a ubiquitous system of control that admits of no alternative and easily co-opts all contrary positions.

Schelsky and Marcuse, in basically similar ways, champion the technocracy thesis which holds that for the modern industrial society, the inherent requirements of science and technology replace democratic-political decisions. Habermas was to deal with this topic in his 1968 essay *Technik und Wissenschaft als Ideologie* (*5.12*). Technocracy results in the denial of certain basic rights because technological factors are inappropriately used to provide an ideological court of appeals for the retention of objectively outdated power structures. Marcuse and Habermas agree in their stated goal which is the awakening of the people from their unreflective and uncritical attitude, and the opening of their eyes to their real interests (their "true consciousness"). They differ, however, as to political strategy. Marcuse advocates revolutionary measures, while Habermas sees the solution in a reorientation of standards and values toward those which espouse a meaningful existence free from domination and pressure to achieve.

Of late a discussion of the "internal" or "external" determinants of scientific and technological progress has gained some prominence. In part, this discussion is concerned with the social aspect of technology. (For an overview, cf. Peter Weingart, *5.37*). One of the issues involved is a question in the *history of science*. Is the present state of science and technology the result of the unfolding of internally generated problems or rather based on external social demands? Obviously, evidence can be cited for either point of view (Wolfgang Krohn, *5.22*, pp. 30–43). The final answer to this question is not merely of theoretical interest; it also has practical consequences for *research policy*. If, in fact, high importance is attributed to external determinants, then, through appropriate controls, it should be possible to channel future development in specific directions as deemed desirable. Since, however, the potential for creative research must be maintained, there will always be definite limits to any exclusively practical orientation of science.

5. THE EARTH AS A SYSTEM

The 1972 Club of Rome study *The Limits to Growth* (Donella Meadows et al., *5.28*) explores the consequences mankind must face if established technological and economic trends are continued. On the basis of past trends, and using realistic assumptions about the projected increase in world population, consumption of raw materials and energy, and the damage to the environment by harmful substances, several model cases are computed. Even in the most optimistic case, it becomes apparent that a catastrophe is unavoidable in the foreseeable future if present lifestyles do not change. While these calculations have by no means gone unchallenged, and have since been modified and supplemented, their real significance lies in the basic soundness of the approach toward problems of the modern technological world. The earth is regarded here as *one system*, and what is investigated is the exponentially increasing rate of growth of relevant processes. It is a foregone conclusion, for example, that total world population, pollution of the environment, and consumption of raw materials and energy cannot increase without limits. Even with the use of new technologies, such as increased food production, more appropriate utilization of synthetic materials, nuclear fusion, or exploitation of the ocean resources, there is still a limit, and this limit, the authors say, will be reached within the next one hundred years. Therefore, they consider it crucial "to alter these growth trends and to establish a condition of ecological and economic stability" as fast as possible (24).

A decidedly gloomy picture of mankind's future development is painted by Herbert Gruhl in *Ein Planet wird geplündert (5.11)*. He points out that in earlier times, man lived for thousands of years practically in a stable environment. By contrast, within the past two hundred years, the face of the earth has undergone fundamental changes. Through irresponsible and ever-increasing pillaging of the wealth of nature, and through destruction of the ecological balance, mankind has reached a point at which production doubles about every fifteen years. This grandiose triumph of industrial technology must necessarily end in catastrophe. The continuation of the prevailing mode of economic behavior is, in fact, tantamount to "programmed suicide" (293, 220). The only possible remedy is to free ourselves from the idea that material resources are unlimited. Present-day economics and politics consider only the immediate future. Human survival, however, requires us now to begin considering the inescapable consequences of our actions.

It is particularly relevant in this regard to distinguish between renewable and non-renewable basic resources, and to consider recycling on as broad a scale as possible. Furthermore, raw materials and energy must be frugally used, and in every case, ecological considerations must be given priority over economic ones (23, 293).

For Ivan Illich, the solution to the problems posed by technological development lies in consistent "self-limitation" (5.17). On the basis of his Third World experiences (Mexico), he vehemently demands a critical assessment of technological innovations. Science as we know it today, with its bias toward technological progress, reduced costs of production, and the administrative control of people, must, he suggests, be replaced by "radical research" aimed at serving two purposes. In the first place, it should establish the criteria for desirable tools. In particular, it needs to determine where the threshold of harmfulness of a technological procedure lies. "Convivial" tools such as the bicycle or treadle sewing machine do not require a mechanical source of energy and can thus be used by each individual to achieve personal goals. By contrast, devices such as jet planes or industrial assembly lines have a "manipulative" character; man serves the machine (52–53). Secondly, radical research should seek to "maximize the freedom of every individual" and to "safeguard creative autonomy and a just system of distribution" (142). At the same time, however, Illich maintains that effective limitation of population growth, environmental pollution growth, waste of resources, and over-industrialization cannot be achieved by political institutions or technological and administrative measures. This can only be accomplished by the autonomous establishment of goals by individuals themselves. Even at this, it still remains an open question as to how these (divergent) goals can be determined in a mass society and established without an institutionalized procedure.

The Marxist theoretician Wolfgang Harich, in sharp contrast, argues for an ascetic system of distribution imposed by an efficiently organized state that is equipped with dictatorial powers. Such a state must reverse the already advanced destruction of the environment and avert the threat of a world-wide crisis of food supplies. In his book *Kommunismus ohne Wachstum* (*5.13*), published in West Germany in 1975, he demands "man's collective ownership of all the planet's means of production" (170), and further calls for a struggle against all needs that are contrary to nature or are intrinsically unsatisfiable for all members of society in the same manner (180, 181). Such present-day needs as privately-owned cars, fashionable clothes and a week-end cabin at the lake can no longer be justified. "Mixed forests are now more important than women's fashion attire" (148). The severity of his theses places Harich in opposition to the generally accepted technological optimism of the Eastern Bloc, according to which communism will bring unlimited development of productive forces and affluence for the entire society. While the orthodox view holds that only communism can guarantee the full *development* of technology, it is Harich's opinion that only the communist social system is in a position to effect a *limitation* of technology.

6. THE PROBLEM OF DIVERSITY OF APPROACHES

This overview, which is by no means exhaustive, already shows the many ways in which technology can be interpreted and evaluated. In the course of historical development, there has, however, been a definite shift in emphasis. In the face of the manifold problems posed by the growth and spread of technology, philosophical interest has in recent times addressed issues pertaining to the practical mastery of technology rather than those that are basically theoretical. Nevertheless, it must be borne in mind that any deeper understanding of technology, which goes beyond mere description of observable facts, presupposes comprehensive and fundamental philosophical analysis and reflection.

It is, however, precisely such works that are lacking. Strangely enough, technology has by no means yet received the attention in philosophical literature which is commensurate with its actual significance. In the course of the history of philosophy, a variety of topics have received attention within the context of a particular school of thought, in which different scholars pursue a common approach and systematically elaborate some central problem: the philosophical foundations of theology – scholasticism; the significance of reason – rationalism; the role of sense perception – empiricism; the question of personal existence – existentialism; the structure of scientific knowledge – analytical philosophy of science. Investigation of the philosophical problems connected with technology in terms of some particular school of thought (with the exception of Marxism) has so far been of a rather rudimentary character. As a consequence, there exists, in this area, no generally accepted theoretical frame of reference or inventory of methodological tools to which one can resort in any particular investigation (cf. Kranzberg and Davenport, *5.21*, 11).

It should come as no surprise, then, that the present philosophy of technology exhibits a broad spectrum of differing approaches. While many investigations focusing on a single aspect of technology may in fact be compatible with one another, the heterogeneous character of the enterprise as a whole remains undeniable. Within the literature, one finds diametrically opposed assessments of such issues as the motives that led to modern technology, the nature of the present situation, and, even more so, the question of future development. Much of the literature is tied to some particular scientific discipline or some already well-established speciality within philosophy. For example, philosophy of technology has been discussed in the context of cybernetics, sociology, ethics, philosophy of history, and metaphysics. Because they derive their substance from an outside frame of reference, these investigations lack a unified, comprehensive conception of the philosophy of technology. Such variations in *concepts* and *methods* are paralleled by marked differences in *content*. Within the philosophy of technology, one can find extremes such as the celebration of technology as savior, on the one hand, or its condemnation as the devil incarnate, on the other.

These differences in approach were apparent at recent conferences the main theme of which was the philosophical problems of technology. (Cf. the Xth German Congress for Philosophy, Kiel, 1972, *0.6*; the XVth World Congress of Philosophy, Varna, Bulgaria, 1973, *0.7*). If one pages through Mitcham and Mackey's comprehensive, annotated bibliography of the philosophy of technology (*0.1*), this impression is reinforced. It therefore comes as no surprise that within the field of philosophy of technology anthologies are more common than in other areas. These contain a variety of perspectives concerning the complex, general phenomenon of technology (cf. for this Hans Sachsse *0.3*).

Procedurally this study adopts the *analytical* approach to the philosophy of technology. This is not intended to prejudge substantive questions. The aim of this work is to present a philosophical analysis of technology which takes into account the historical and systemic aspects of technological development, provides a thematically ordered overview of the pertinent problems and basic solutions, and, at the same time, makes a contribution of its own to the relevant issues. The range of themes to be investigated in this study becomes clearer if one attempts to distinguish this philosophy of technology from similar studies. It is neither (1) a metatheory of the engineering sciences, nor (2) part of a separate discipline such as sociology or history, nor (3) can it be reduced to a single philosophical speciality. It is closely related to epistemology, social philosophy, philosophical anthropology, philosophy of history, and metaphysics without being subsumed by any one of them.

The first characteristic of technology is that it is always a factually given phenomenon. Its actual features cannot be deductively derived from a contemplation of its logical, timeless essence while disregarding concrete, empirical evidence. To avoid arbitrary and nonconvincing speculations, philosophical analysis and reflection must be based on the contingent facts. Proceeding from such a point of departure, generalizations do become possible. In what follows, this 'inductive' approach will usually be preferred, although the legitimate philosophical interest in the 'deductive' method will be given due consideration. Since the pertinent phenomena are complex and intertwined, the chosen approach excludes *a priori* a strictly linear development of the argument. It will, therefore, be necessary to return to the same problems in different contexts with accompanying shifts in emphasis. This procedure, it is hoped, will do greater justice to the problems under discussion than would theoretical consistency and smoothness bought at the price of oversimplification.

CHAPTER II

DIFFERING VERSIONS OF THE CONCEPT OF 'TECHNOLOGY'

1. PROBLEMS OF DEFINITION

At first sight, the meaning of 'technology' seems completely clear. Because we encounter technological apparatuses, devices, and procedures on every hand we have come to accept them as 'second nature'. As soon, however, as we are called upon to supply a clear and unequivocal definition of the concept of technology, difficulties arise. The situation here is similar to that of other familiar concepts which are also of a highly generalized character. Whereas each of us thinks he knows what is meant by 'science', 'politics', or 'society', agreement on a precise definition is difficult to obtain. In fact, given the manifold determinants of technology, it is unreasonable to expect universal agreement upon any one definition.

Nevertheless, in the present context it would be a mistake to do without any definition at all, along with, perhaps, the half jesting, half resigned observation that 'definitions only confuse'. While granting that a degree of imprecision is inherent in the subject, it is nevertheless possible, through juxtaposition and comparison of various proposed definitions, to make mere intuitive understanding more explicit and to render it accessible to scrutiny. Because of the elements of ambiguity and vagueness inherently connected with 'technology' one is forced, in order to arrive at a precise demarcation, to make certain assumptions within the range of subjective opinions. Because 'technology' is now in common parlance we cannot, on the one hand, simply introduce a completey novel stipulation. What is meaningful and necessary, on the other hand, is an appropriate differentiation of the semantic elements contained in the various existing definitions. The very fact of the widespread use of the word can be taken as an indication that 'technology in its broad meaning, is concerned with an intricate and manifold complex of phenomena which are, by their very nature, closely related.

2. HISTORICAL AND SYSTEMATIC ANALYSIS

Nietzsche has appropriately remarked, "only that which has no history can be defined" (*4.22*a, 212). Strictly speaking, this would imply that an ahistorical definition of 'technology' is impossible. Since technology is a historical phenomenon it could be adequately conceptualized only within its particular historical context. Indeed, the production and use of technological objects always has a concrete historical setting which is itself embedded in a superordinate historical context. The complexity of particular technological systems, the manner of their production, and the actual consequences of technological activity are so different in different times as to incline one to speak of "family resemblances" (Wittgenstein) among different phenomena rather than instances of one phenomenon.

Such a view, which particularly emphasizes the uniqueness of each technological process, however, does justice to only one side of the phenomenon. As with other fields as well, the individual historical stages of technology are parts of a temporal continuum. The development of metal processing through various stages, along with the specialized artisan technology of advanced cultures up to the modern combination of engineering art and modern science, which has spread throughout the world, has been a continuous process since the Stone Age. Seen as a whole, this development may be characterized as one in which the complexity of the procedures employed and the performance efficiencies of technological artifacts continually increase. On a less general level, however, when one focuses on details, a host of temporal and geographic distinctions are required. In particular, one needs to account for the phenomena of multiple discoveries, diffusion of inventions, and even the loss of technological skills and insights. Any corrections of this kind, however, are only modifications of the general trend which is towards ever increasing technological perfection.

Granted the fact of historical diversity, technological procedures, nevertheless, always exhibit *common* features. While it is their concrete embodiment which determines the character of any particular era, such features make possible a supra-historical structural description of technology. In the first place, technological action is always tied to ideological presuppositions, such as a certain world view and a corresponding state of technical knowledge. To these theoretical prerequisites we must add the element of commitment (motivation) required to channel technological potential in a certain direction. Furthermore, concrete execution of the technological process depends on certain material prerequisites such as raw materials and tools along with appropriate forms of organization (specialization, division of labor, and processes of economic exchange). Lastly, beyond the immediate, intended results, we must account for the secondary impacts of technological activity on man himself and on the environment. All of these factors vary with the particular historical context. Their specific form determines the character of each concrete situation, a specific image of technology itself, a corresponding state of technological development, and the role that technology is to play in social and cultural life. The foregoing is illustrative of the fact that the philosophical problems of technology can be approached in two ways. One is historical, the other systematic. Neglect of either approach amounts to an inappropriate curtailment of the problem. Consequently, in what follows I shall first give a rough outline of the historical stages of technology, and subsequently treat systematic attempts to define 'technology'.

3. PERIODS IN THE HISTORY OF TECHNOLOGY

Man may be characterized by his biological deficiencies (Gehlen). In order to hold his own when confronted by nature he has always had to resort to certain technologies in the face of nature's demands. Even his most basic needs, food, clothing, and shelter, require the production and use of appropriate tools. Health care, transportation, and defense warfare have also, at all times, depended upon technological devices. Additionally, from the beginning, technology has served as a means of religious and artistic expression. Each technological measure both reflects a concrete historical situation and, in turn, provides the foundation for future technological activity. Each individual person, and in a broader sense, each historical era, is a link in a progressive process, the concrete situations of which are characterized by the specific way in which tradition and heritage are utilized.

The advanced state of modern technology tempts us to underestimate the technological accomplishments of earlier times. However, when one considers the fact that even in our own time the discovery of entirely new principles is relatively rare and that technological progress is for the most part a process of gradual development of existing procedures, then it becomes clear how important such inventions as the wedge, drill, needle, lever, plow, and wheel must have been. One might even go so far as to consider prehistorical and early historical times as the most important periods in the developments of human technology (Hans Rumpf *3.9*, 90). The very fact that entire epochs of human prehistory have been characterized as 'Stone Age', 'Bronze Age', 'Iron Age', serves to underscore the preeminent importance of technology.

Ortega y Gasset sees the decisive principle for dividing the history of technology into distinct eras to be the particular concept of technology which prevailed at the time. He distinguishes three such distinct concepts: (1) the technology of chance; (2) the technology of craftsmanship; and (3) the technology of engineering science. The technology of chance is typical of prehistoric man and is illustrated today by the contemporary primitive, (1). In this stage technology remains totally embedded in an unreflective, animalistic flow of natural life. There are no skilled craftsmen, and invention is a matter of accident rather than intention. Classical antiquity and the Middle Ages are characterized by the technology of craftsmanship, (2). The variety and complexity of technological procedures which are developed become so great that a division of labor results. Knowledge and practice of a particular process remains restricted to only certain artisans. Technology is integrated into a traditional system of skills or 'arts' — the institutional framework of guilds, for example — comprising by and large an unchanging repertoire of know-how. Our time, by contrast, is entirely dominated by the technician or engineer, (3). What is decisively different here is the fact that tools, such as machines and other apparatus, have acquired a measure of autonomy; they are no longer operated by people but are rather, as it were, detached from them. In the case of the power loom, for example, man only *controls* the work process; he does not supply the *muscle power* for its operation. While human capacity for physical work is quite limited, the use of self-activated machines opens up an almost unlimited range of possibilities (*1.25*, 88–103).

A somewhat different approach to the stages of technological development is provided by Scheler. He distinguishes four fundamental transitions: (1) from magical technology to the positive technology of weapons and tools; (2) from the cultivation by hoe of matriarchal cultures to agriculture (use of the plow) and cattle raising; (3) from tradition-based, manually operated tools, to the science-based, and efficient self-powered engine; (4) from the system of early capitalism to the use of coal as an energy source. (*4.26*, 137–8). Gehlen holds, on the other hand, that there have actually been only two historical junctures of primary importance.

These are (1) the "neolithic revolution" in which mankind made the transition from a life of nomadic hunting to a sedentary one of agriculture and cattle raising, and (2) the changeover to "machine culture" of the Industrial Revolution. Both changes have had very far-reaching consequences. Just as settling down led to large concentrations of population along with differentiations in wealth, domination, division of labor, and the emergence of local deities with their temples and cults, so, too, the transition to the Industrial Age has resulted in fundamental changes, the consequences of which, even today, cannot be fully assessed (*5.8*, 25–46).

Ladislav Tondl has arrived at a systematic categorization of ideal types of technological devices which, in a rough way, also corresponds to the phases of the history of technology. It is based on a comparison of the share of work done by man himself with that part performed by technological devices. As technology has developed, the efficiency of technological apparatus has increased to the same degree that the amount of required human work has decreased. The three ideal types of technological devices are: (1) the tool, (2) the machine (in the classical sense), and, (3) the automaton.

(1) A tool such as a knife, an ax, screwdriver, or a chisel is activated by muscle power. With the help of the tool, man acts upon the object of his work and modifies it according to his goals. Man controls the tool, and is also, the source of information. (2) In the classical sense, machines are devices powered by non-human sources of energy — draft animals, wind, or water. Introduction of the heat engine (e.g., steam, internal combustion) resulted in a significant increase in efficiency. All classical machines, which consist of an energy source, transmission mechanism, and a particular operational apparatus, nevertheless, are still controlled by man. (3) With the automated machine, the application of cybernetic principles of control and regulation renders largely superfluous even the control function supplied by man. We are not

THE CONCEPT OF 'TECHNOLOGY' 29

dealing here with perfect automata since control and decision-making is still determined by a program which is of human design. Experiments with machines which learn and with self-organizing systems, however, have as their goal the further elimination of the human factor and its replacement by automatic processes (*2.5*, 9—11).

Ribeiro's attempt to describe the basic outline of human prehistoric and historic development as the result of technological processes is noteworthy, in this regard. He considers "the concepts of technological revolution as the basic causal factor, sociocultural formations as theoretical models of cultural response to these revolutions, and civilizations as concrete historical entities that are crystallized out of the formations" (*4.24*, 29). Ribeiro accepts the three generally recognized "cultural revolutions": agrarian, urban, and industrial. Between the latter two he introduces additional stages: irrigation, flock tending, and mercantile revolutions. Finally, beyond the Industrial Revolution (introduction of power machines) he distinguishes the current "Thermonuclear Revolution" (automation, computer technology, utilization of nuclear energy), which other authors have called the "Second Industrial Revolution" (36—43).

The value of such a classification is obvious. It provides quite an appropriate and informative way of systematically considering long stretches of time. This is not at all surprising, since the outward appearance of an era is necessarily shaped by the technologies involved. Since the concrete physical foundation for the preservation of life and for development of culture can be secured only through appropriate technological procedures, technology indeed plays a key role in the historical process. This is particularly apparent in the case of armed conflict, where, as a rule, the more 'efficient' technology tips the scale in the struggle for survival.

Ribeiro, however, goes beyond the assertion that technology plays a key role in the historical process to make the stronger claim that, from the long-range perspective, technology is the decisive factor. Two counterarguments in particular weigh against this thesis. First, political, social, cultural, and religious life in any given society cannot be derived in a straightforward way from the accompanying state of technology. During the Middle Ages, for example, the general levels of technology in China, the Islamic countries, and Europe were quite comparable, while, at the same time, the characteristics of individual as well as collective life were quite disparate. One could, of course, ignore such particular variations, in the name of long-range analysis. The price to be paid, however, would be that of an abstract and comparatively impoverished characterization, in which the concrete diversity of the historical process is reduced to extant signs of technological process.

It is not only questionable as to whether Ribeiro has provided an adequate description of the phenomena; a second counterargument challenges his claims about technological causality. The claim that technology is the sole determinant of all historical processes can hardly be verifed empirically. The primary data available to the historian give evidence, after all, only of *simultaneity* between certain technological, social, and cultural phenomena. The question as to which of these factors is causally prior must be left open to further investigation, which must itself entertain the possibility of mutual causality. Jakob Burckhardt's *Force and Freedom; Reflections on History* (4.5), for example, demonstrates that a competent world history can be written with virtually no reference to technology. For him, the historical process is determined by the interaction of three factors: the state, religion, and culture (147–213). Burckhardt's elaborations, it is true, date back more than a hundred years. Because its present-day role is so obvious to us, there is a tendency in retrospect, to overemphasize the role of technology in our interpretation of history. Technology, after all, occupied a rather less important place in the 19th century than in the present. It is, however, no mere coincidence that, in Ribeiro's case, it is an anthropologist trained in research on the American natives who considers technology to be the determining factor. Judging by appearances, technology was as important for prehistorical and so-called primitive cultures as it is for our times.

4. SEMANTIC VARIATIONS OF THE CONCEPT OF 'TECHNOLOGY'

Because so many different forms of technology have occurred in the course of *historical* development, no one definition of it can do justice to every particular case. There is, additionally, a *systematic* ambiguity hindering precise definition. (To avoid unnecessary complication, we consider here only present-day usage; for the historical development of the concept of 'technology', cf. the detailed study by Winfried Seibicke *2.2*).

In the narrow sense, we understand 'technology' as 'technique' – a certain *procedure*. This refers in the simplest case to a learnable skill, such as the technique of driving a car, playing the piano, or skating. Complex processes which are the result of collective action, however, are also rule-governed and goal-oriented and thus dependent on a certain technique. Here, we tend to speak of the technology of mass advertising, the technology of energy production, or the technology of automobile manufacturing. According to this broader, less well-defined meaning, we subsume also all *objects* connected with such procedures under 'technology'.

Such objects are *used* in technological processes (e.g., instruments, tools, machines) or are *produced* as the result of such processes (e.g., semi-finished or finished products). Since the efficiency of technological actions largely depends on the type of tools used and since the tools themselves are the product of other technological processes, one common simplification is to treat as one both technological procedures and objects. The preceding, however, is only a summary characterization. Upon closer scrutiny, from the perspective of technology as efficient, goal-oriented activity, two aspects emerge: technology as procedural *knowledge* and as *actual execution*. These aspects are, however, closely interrelated, since, in the case of complex technologies, mere rules of thumb do not suffice. Rather, prior theoretical knowledge becomes the indispensable condition for successful practical action.

The need for theoretical knowledge moves technology closer to *science*. The rise of modern technology is, indeed, closely tied to the development of the engineering sciences and related scientific disciplines. In many areas, such as communications technology or data processing, the present state of technology would be quite inconceivable without the corresponding theoretical foundations. On the other hand, there are also broad areas of technology, such as building construction, mining, and machine design, which are based far more on practically proven traditional rules of thumb than on theory-based scientific knowledge.

Complex technological activities are always *social action processes*, involving small or large groups of participants. The present high level of efficiency is based on coordinated cooperation within a framework of specialization and division-of-labor. This holds as true for technological research as it does for production and product utilization. While many technological products are consumer items, technology as a whole is the result of social action. It becomes immediately apparent how technology is embedded in its social context if one broadens the scope of investigation. Consideration of the automobile, for example, must also include the production and transportation of raw materials and gasoline, maintenance and repair, road construction, and environmental impacts. Beyond the immediate satisfaction of needs — the transportation service provided by the automobile — the effects of technology always concern the entire society, since the private sphere, the world of work, the social structure, institutional divisions and value attitudes are formed, directly or indirectly, through the technological lifestyle.

If one does not focus on the action processes but rather on the material substratum upon which the respective actions are carried out, then technology is seen to be the result of purposefully caused *natural processes*. Such processes are artificial inasmuch as technological objects and processes do not occur in nature untouched by man; they are natural, however, inasmuch as they are subject to the natural laws of the material world. The Greek word *techne* reflects this duality, referring both to the artifice, the cunning device, as well as to creative and ingenious (ingenieur) human technical skills. From this point of view the inventor who himself typified previous stages of technological development, resembles the artist. During the Renaissance, it was possible for Leonardo da Vinci, for example, to combine both capacities.

It is immediately apparent that the definition of 'technology' depends largely on which meaning variation – action process or material substratum – or other aspect is focused on.

5. ATTEMPTS AT DEFINITION

In view of the multitude of relevant aspects mentioned, we are faced with the choice of either accepting a short and vague formulation or of thematicizing the individual aspects one by one. In this latter case, however, one has already left a discussion of the framework of a clear conceptual definition for one of details. This situation can be illustrated with the proposed definitions of various authors. (Cf. the survey by Lenk-Ropohl *1.20*, 110–111).

A broad characterization of technology in the metaphysical tradition is given by Heinrich Beck who considers technology to be the "transformation of nature through the intellect . . . . Man transforms and changes the inorganic, organic, and his own psychological and intellectual nature (or the corresponding natural processes) according to his goals, in proportion to his understanding of the natural laws" (*1.6*, 29–31). A similar definition provided by Eyth is "*technology is everything that gives a corporeal form to human will*. Since human will nearly coincides with the human spirit, which itself is comprised of an infinity of manifestations as well as possibilities of life, technology too, despite its dependence on the material world, has something of the limitlessness of the world of pure spirit" (*3.4*, 3–4). Friedrich Dessauer, too, focuses on the idea of creative transformations. After enumerating numerous proposed definitions, he himself offers the following definition of the nature of technology: "Technology is reality derived from ideas, through purposeful forming and processing of natural resources" (*1.11*, 234).

Other authors, however, see the essence of technology not in the fact of *human activity*, but in the *methodological character* of the procedures employed. Thus, Hans Sachsse emphasizes that the success of technological measures is achieved through subdivision into different intermediate stages and into specialized functions – a procedure which although actually a detour, leads most quickly to the intended goal (*5.33*, 51–53). In the same vein, Fr. von Gottl-Ottlilienfeld defines "technology in the *subjective* sense as the art of the *right way to an end* . . . and in the *objective* sense in the *refined totality of procedures and instruments within a defined specified area of human activity*." As he further explains, "that way is always the most reasonable one *which, calculated by the unit of success, requires the least expenditure*. For it is only logical that this is also the way which, calculated by the unit of expenditure, provides the maximum success" (*2.1*, 207, 211). As a matter of fact, the continuous technological progress since the Industrial Revolution has been based on an increase in the efficiency of technological procedures (Henryk Skolimowski *2.3*, 74–85).

The class of systematically perfected techniques or procedures

THE CONCEPT OF 'TECHNOLOGY' 35

is quite broad. Strictly speaking since every action that is consciously and purposefully executed follows a methodological pattern, however rudimentary, one would have to classify all purposeful (individual as well as social) action as technological, we have to distinguish this very *broad* definition of 'technology' from the *narrower* sense of the word which restricts 'technological activity' to the performance of the engineer. Von Gottl-Ottlilienfeld speaks in this connection of "material technology", which aims, in contrast to other individual, social, and intellectual procedures, at the *domination of nature* through transformation of the *outside material world* (*2.1*, 207). The definition by Tondl points in the same direction. For him technology is "everything that man interposes between himself as subject and the objective world for the purpose of changing certain characteristics of the world so as to make possible the achievement of certain goals." Tuchel gives a similar, but more concrete definition. "Technology is the general term for all objects, procedures, and systems, which, on the basis of creative construction are produced for the fulfillment of individual and social needs; which through defined functions, serve certain purposes; and in their totality change the world" (*2.6*, 582).

In the following, unless otherwise noted, 'technology' will refer to material technology which is based on action according to the engineering sciences and on scientific knowledge. This definition comes most closely to the usual understanding. Furthermore, it is the only practical approach, for if one were to actually include not only material technology for consideration, but *all* methodical procedures, then one would have to treat such topics as the construction of legal systems, the methods of political action, or the principles of warfare. Such an investigation would necessarily move the specific problems of engineering technology into the background. While it is certainly possible to treat together the different types of methodically based and efficiently executed action processes (cf., Kotarbiński, *Praxiology*, *3.6*), such an investigation is only loosely connected with the specific questions of the philosophy of technology. (Regrettably, the desired terminological distinctions between concrete material technology and general methodological procedures cannot be accomplished by equating 'engineering technology' with 'technology'. Usage is not at all uniform. An overview of proposed definitions of technology from a Marxist perspective may be found in Kurt Teßman, *2.4*).

CHAPTER III

METHODOLOGICAL ANALYSIS

1. THE DETERMINANTS OF TECHNOLOGICAL DEVELOPMENT

The determinants involved in the production and use of technological artifacts can be identified in a methodological investigation. We are not concerned here with the parameters of concrete spatial and temporal instances of technological action processes, but with the general structure of applied methods, which can then be analyzed into elements and process phases. With this turn, attention appears to shift from practical questions to abstract theoretical problems. Nothing, however, is more practical than a good theory. Thus, methodological analysis supplies not only fundamental — and thereby also philosophical — insights into the structure of technological action. The practical efficiency of modern technology, too, is largely based on a systematic perfecting of procedures and thus, in a broad sense, on methodological analyses. In the artisan tradition action is governed by the rules of experience, which are based in concrete reality. By contrast, in the engineering sciences the range of the technologically possible is systematically expanded by keeping a conscious distance from immediately given circumstances, and relying rather on general theoretical considerations (Hübner 3.5). It is evident from appearances too that the complicated apparatuses and systems of modern technology represent by no means direct solutions to problems. They reflect instead intentionally taken detours through theory formation and methodological reflection.

In what follows, three different models of the development of technology are examined. Each model will emphasize different factors, driving forces, or determinants of technological development. The trick is to focus only on the variables of interest in each case, keeping all other variables relevant to the subject matter outside one's attention, or considered with a *ceteris paribus* reservation, constant and therefore negligible.

(1) The starting point for the *decision-theory model* is the statement that given the multitude of possibilities of technological action, a certain selection process must take place to determine which alternatives will subsequently be realized. Three difficulties in particular stand in the way of a simple and clear description of this complex decision process. In the first place, its *structure* is not at all self-evident since a wide range of individuals, groups, and social institutions, each participating to a different degree, are involved in what is a complex and intertwined process. Secondly, decisions are, as a rule, made in an unreflective and intuitive rather than a *conscious* manner. Thirdly, therefore, the (diverging) *value concepts* and *decision criteria* applied are frequently only identifiable by means of theoretical reconstruction. Despite these difficulties, however, formulation of a decision-theoretical model – one that can then be expanded and improved – is the only alternative to a vague general understanding that is lacking in sufficient detail.

The design and construction of technological systems is the business of the *engineer*, who thus could be considered as the authority for decisions in technological actions. Indeed, his proper competence begins and ends with the making available of technological means for the solution of a particular problem. Should he additionally claim the right to determine the goals of technology, he would be asking more than a democratic community is prepared to grant. Nevertheless, the engineer, in his capacity as expert in a high-technology society, has a particular obligation to point out the negative impacts to be expected from specific technologies.

General specifications for a technological project are usually prescribed for the engineer by his *superiors*. In a private enterprise system, for example, this means the management, and in a planned economy, the central planning bureau. In both cases the decision is tied to the anticipated attitudes and responses of buyers and consumers. Products are, after all, for use. From this point of view, it is the economy that decides. For the most part, this generalization holds although it needs qualification and correction because of certain complicating factors: reduction of supply, the presence of market-dominating enterprises, or consumption influenced by advertising.

Economic life, however, is in turn subject to intervention and regulation by *political institutions*. This brings another level of decision-making into play, one, in fact, which in actual practice often deviates from the ideal of representative democracy in that it is susceptible to special interest groups and bodies of experts and to a public opinion that, to some extent at least, can be manipulated. Given this superposition of decision-making structures along with the differing goals and interests of the various decision makers, it thus becomes apparent that one cannot speak of a clearly delineated and uniquely definable selection process. Yet, it is equally obvious that, in the last analysis, the trends which emerge as a result of this complex process, rather than being naturally given, reflect the choice behavior and value preferences of the individuals involved.

(2) Those who actually make the decisions, however, do not act in monadic isolation. They themselves are embedded in a complex network of social relationships and an impersonal current of events which they co-determine without being able to escape its powerful influence. Given the fact of a highly differentiated specialization based on division of labor, and in view of the universal expansion of the technological life style, the technology which has evolved has today become a decisive force on which the future fate of

mankind largely depends. This situation can be adequately grasped only within a *social theory model* of technological development.

In such a model one can, in principle, distinguish between 'external' and 'internal' prerequisites of technological action. *External* prerequisites include, for example, demographic factors, social differentiation and stratification of society, as well as technology-related exchange and distribution processes. Production of commodities and services, for example, can only begin if the necessary raw materials and apparatuses have been previously prepared. Final products, in turn, require distribution in order to reach their prospective users. The entire process of production, distribution, and utilization, based on the division of labor, is also always dependent on the political system by means of which technological activity is channeled in a specific direction.

The *internal* prerequisites upon which technology depends are less easily identified, but are not for that reason any less important. Man is not, after all, merely a *product of nature*, reacting to external stimuli according to certain laws and instincts, but also equally a *product of culture* defining himself in freedom. Every action that is premeditated and purposeful reflects, in however implicit or rudimentary a form, values and objectives it is intended to realize. This is especially true of project-oriented technology. Yet immediate technological tasks are neither autonomous nor transtemporal. Those concrete goals gradually emerge from the permanent interaction of technological potential and the intellectual and cultural milieu.

At first sight, consideration of the external and internal determinants of technology appears quite unproblematic; no one contests the relevance of either type. Furthermore, one is not likely to dispute the claim that concrete physical conditions and intellectual motives are somehow connected. The real problem lies in how this interdependence is to be described in a conceptually adequate way, that is, without prematurely declaring that either the 'materialistic' or the 'idealistic' conditions are the only relevant ones. Each of these extreme positions only appears to be simple, for in either case one is forced to derive the complementary phenomena from the one regarded as decisive. Additionally, besides the systematic interdependencies, one also needs to take into account the prior history of developments of the technology in question.

(3) The *action theory model* ties in directly with the relevant action processes themselves, and in this respect, comes closest to the actual course of events. The starting point here is the goal-oriented subjugation and appropriation of nature by man. This process is always connected with the technological, social, and philosophical prerequisites that are required for the particular form of technology to arise at all. The *technological* prerequisites include initial products (raw materials and energy), concrete material aids (instruments, machines, apparatuses), as well as, in a broader sense, a particular state of technological knowledge and ability. These factors pertain directly to the material world, transformation of which yields the various technological artifacts. The counterpart of nature, the object of work, is man, the subject who acts. Accordingly, *social* prerequisites involve, in the first place, the individuals who produce, distribute, and use the products of technology. They further comprise concrete work processes and, in addition, organization principles (division of labor, specialization) and economic exchange processes.

The constellation of technological and social conditions provides only the potential for technological activity. In any particular case the decision remains open concerning its actual realization. As a further determinant we must add appropriate motivation. This reflects a specific order of values, which in turn defines what is considered to be worth striving for. In the last analysis, such an evaluation can always be reduced to *philosophical premises*, i.e., to certain views of human existence and of nature. It is only on the basis of this evaluation, that the actual decision situation can acquire an imperative character, and thus induce a specific kind of technological action.

2. THE RANGE OF ACTION

Technological action can neither be arbitrarily chosen nor executed. As with other types of action, the range of possibilities is specifically limited. The parameters of this range reflect the various prerequisites that must be satisfied in order that the envisaged measure can be meaningfully planned and successfully executed. In the following, the various relevant limitations are examined in more detail. The basic approach will be to start with the most general conditions and, by means of the appropriate distinctions, illustrate how the range of realizable technology is progressively narrowed, as additional factors are considered.

Technology consists in man's exerting influence on his physical environment. Accordingly, there are two roads open here for investigating particular action possibilities. One can begin either by focusing on the limitations pertaining to the *material world*, the object to be worked on; or by identifying those restrictions concerning man, the *active subject*, upon whom the process rests. Because it is much simpler to describe the processes in the material world that are caused by technological activity than to account for the psychological and social determinants of technology, we shall begin by exploring those restrictions arising in connection with the material world.

Intervention into the process of nature is only possible within the range marked by the laws of the physical world. This is not yet the most general type of restriction, however. For if one tries to formulate the applicable natural laws in the language of mathematics, it then becomes evident that a non-contradictory mathematical calculus is required for the description of actual or potential states of affairs. Thus (1) *logical consistency* forms the most general restriction imposed on any technological action. In terms of the concrete execution of technological projects this means that it is impossible to realize objects or states of the physical world which simultaneously possess and lack a certain well-defined property.

Of course, by far not all logically consistent processes and phenomena of which one might conceive actually exist in the material world. Empirical, experimental investigations are required to determine which mathematical calculi describe actual physical processes. In the area of scientific disciplines this is evident in the transition from pure mathematics to theoretical physics. Hence beyond the demands of logical consistency technological action is further limited by (2) the *laws of nature*. In particular, these lay down the boundaries for technological processes which one can achieve in the best of cases, but which one may never exceed. An instructive example of this is offered by the various energy conservation theorems. One theorem states, for example, that it is impossible to build a *perpetuum mobile*. Similarly the second law of thermodynamics excludes the possibility that any electrical transformer or combustion engine can ever achieve 100 per cent efficiency. Other examples of limiting theorems concern the propagation of physical effects with the velocity of light (Einstein), the measurability of microscopic processes (Heisenberg), and information transfer through a given channel (Shannon) (cf. Tondl *2.5*, 14–16).

The historical development of the sciences and the course of the history of technology show that our knowledge of the laws of nature so far has always been imperfect; however, at least in the modern age, a continuous progress in knowledge can be seen. Therefore, at each historical stage, only a certain percentage of what the (unknown) laws of nature permit in the way of potential technological action can be realized; only those laws of nature that have actually been understood can be exploited and transformed into technological constructions. Consequently (3) the *state of scientific understanding* represents an additional restriction on potential technological action. Thus, for example, radio and television sets or nuclear power stations could be built only after the corresponding scientific principles were known. From the point of view of technological practice, scientific research has the task of expanding the range of actual scientific understanding (3) to a point where it more and more approaches the limits of the laws of nature (2) imposed on the respective natural processes. The real task consists thus in making *subjectively* possible what is *objectively* possible, i.e., what is possible under the aspect of the nature of the physical world.

Scientific knowledge and technological know-how are, however, by no means identical, for particular research results, taken by themselves, are no sufficient basis for concrete technological application. Even if certain experiments have been successfully performed in a physical or chemical laboratory, their economic exploitation at the level of full-scale production is by no means guaranteed. Scientific knowledge does not, for example, provide any information about the construction and dimensioning of the sought-for technological systems nor about the function of any necessary subsystems. Specific problems of engineering science — operational safety, maintaining tolerances, useful life — also lie outside the range of scientific inquiry. Therefore, a further limitation to concrete potential technological action is (4) the *state of technological knowledge and skill.*

So far, the aspects investigated have referred to the structural relations of the material world (logic, laws of nature) and the actual state of intellectual resources (results of scientific research, technological knowledge and skill). All technological activity, however, aims ultimately at producing concrete material objects that are to fulfill a given task. Therefore, beyond the conditions

METHODOLOGICAL ANALYSIS

named, certain physical prerequisites must be met which can be summed up as (5) *material resources*. Among them are raw materials, energy resources, machinery and equipment, as well as — under the aspect of physical work — the necessary manpower. These variables represent an additional limitation in that any particular technological project can be realized only if all the material resources are available in the required quantities. To be sure, in a specific situation, there always exists the possibility of resorting to other sources of energy or to different construction materials; yet, on the whole, the available supply of material resources is, at any given time, just as limited as is the level of scientific understanding and the state of technological knowledge and skill.

The range of potential actions, subject to the aforementioned limitations, is fundamentally incommensurate with the wishes and envisioned goals which technology is expected to fulfill. Wishes and visions are not hampered by such limitations and therefore can practically arbitrarily grow in any direction. Thus a certain selection process always takes place which singles out the projects actually to be realized from among those merely envisioned. Assuming that it is the customer or user who is, in this case, the decisive point of reference, (6) the *receptiveness of the market* is a further limitation on technological activity. The willingness to pay a certain price for a certain technological product can itself be reduced to the *value preferences of the consumer*. For mankind as a whole, these value preferences can be considered to be the result of historically conditioned choice, which is nevertheless, in principle, autonomously made. The individual producer or buyer, however, when producing or buying technological systems, is largely dependent on the value concepts of his society, as reflected in the market potential of different technological products. In this way the standardization of technological products, which initially reflected merely the economics of the production process, comes to exert a social influence. Inexpensive consumer goods and services are possible only in case of standardized serial production, which in turn presupposes uniformity of tastes and predictable buying behavior.

Finally, beyond the economic mechanisms there are also (7) *political* and *legal restrictions* that limit potential technological action. More specifically, these restrictions may be based on goals in areas such as governmental services or international and armament politics. Because of codified legal norms, some technological actions are ruled out despite the existence of a definite market demand. Examples include safety restrictions or bans on certain commodities (arms, electronic surveillance equipment). Measures for preserving the natural environment must also be prescribed by legal injunction, since, as a rule, producers and buyers are interested in their own short-term goals only and do not voluntarily consider the more comprehensive and long term public interest.

By way of summary, the stepwise progression of restrictions on technological action falls into four categories, namely, those concerning

—the structure of the material world (logic, laws of nature),

—intellectual resources (the state of scientific understanding and of technological knowledge and skill),

—material resources (raw materials, energy, machinery, manpower),

—social conditions (market mechanism, political and legal restrictions).

In a first and tentative analysis, one can consider all of these restrictions in terms of a *fixed* set of variables which are characteristic of each concrete historical situation. All concrete technological action can therefore, generally speaking, take place only within the range marked by the limitations given above.

At a deeper level of discrimination, however, it turns out that most of the restrictions mentioned can be *changed* over the course of time. The only exceptions involve logic and the laws of nature, which are in principle removed from human influence. All other restricitions may, within limits, be systematically altered. This is particularly obvious in the case of intellectual resources. Here the task of research consists precisely in expanding the range of the technologically feasible through new insights in such a way that ever more 'efficient' technological systems can be produced. In a similar way one strives to make maximum use of raw materials supplied by nature. The *total amount* of raw material supplies, however, constitutes an understandable upper limit. Social restrictions, which have formed in the course of time, are not externally imposed upon mankind, but are, in the last analysis, chosen by mankind. Nevertheless, any value preferences, just as the state of intellectual and material resources, depend on the preceding historical development. Therefore, there always exists the possibility of channeling the development in a certain direction, with the given conditions as the starting point.

3. THE TRANSFORMATION OF THE MATERIAL WORLD

Technology consists in an alteration of the physical environment in such a way as to produce objects and initiate processes which aim to fulfill quite specific functions. Abstracting from concrete situations, one can distinguish the following decisive factors in this transformation of the material world: the acting subjects; the stated objective; the impact on the physical environment; the instruments or apparatuses utilized; material changes accomplished; and, lastly, the application and utilization of artifacts produced (Kotarbiński *3.6*, 14–38). By means of a discussion of these various points, the characteristics of technological action can be explained in greater detail.

(1) The *acting subject* is, in the most basic sense, an individual person performing a specific technological measure. In all cases of higher complexity, however, the required physical and intellectual

work surmounts the capacity of an individual. While, in the case of artisan technology, individual achievements by master craftsmen and apprentices are typical, modern technology is the result of *collective action*. This is immediately evident in the case of large-scale projects (dam construction, space travel). Nowadays, however, even comparatively simple technological systems utilize the team approach, so that, for example, planning, work preparation, parts production, and maintenance will be performed by different individuals. This intermeshing of many varied individual actions leads to the situation that, for any particular technological project, the acting subject can only be identified as an abstract, collective entity.

If, in addition to production, one also considers the application and utilization of technological systems, then the difficulty of distinguishing between individual and collective aspects becomes even more apparent. The automobile driver, for example, is able to fully utilize his vehicle only after roads, service stations, and repair shops have been provided by others. Certain systems of transportation, communication, and utilities, such as railroads, telephone systems, or natural gas distribution lines, are from their inception designed as uniform and comprehensive technological networks. The combination of specific technological systems with the principles of division and specialization of labor leads to the phenomenon that everything technological is necessarily part of an even larger system. Because of this characteristic and far-reaching interdependence, all individual phenomena are combined to form a comprehensive technological complex into which, it seems, the individual human being is helplessly drawn. Within this anonymous matrix of interrelations, however, all technological objects and processes still, in the last analysis, are originated by concrete individual actions.

(2) The *fundamental objective* of technological systems consists in relieving man from physical work. This becomes immediately

evident, for instance, in the case of power engines and in transportation technology. Additionally, technological devices can increase the capacity of human sensory organs in manifold ways, thereby providing access to phenomena otherwise completely outside our range of experience (electro-magnetic processes, phenomena on the microscopic or macroscopic level). Beyond this, automation and computer technology frees man of even routine intellectual work. The actual intended primary objective of technological processes thus lies in human relief and increased efficiency. Decisive for an understanding of modern technology is the fact that in the course of technological development, new needs continuously appear which are awakened by the expanding potential of technology and which lead to concrete technological projects. 'Technological progress', accepted as the general norm, requires optimal efficiency and rationality in the use of existing technological systems. With this, man has largely relinquished his position as an autonomous subject vis-à-vis technology and has subjected himself to the dictates of real or imagined technological necessities.

(3) The *influence on the physical environment* is experienced, in the simplest case, in tool use, where muscle power is required for the mechanical transformation of a physical object (Kotarbiński *3.6*, 14). In the case of artisan technology, however, the range of nature-modifying action is, in principle, limited. The transition from artisan manufacture to large-scale industrial production, for example, was not possible until new energy sources (coal, natural gas, petroleum), iron as a construction material, and mechanized procedures based on the division of labor were utilized.

This development effected a fundamental change in man's relation to his environment; whereas in earlier times, despite the use of tools, technology for all practical purposes, reflected the *direct* impact of human action on an untouched nature, it now takes place *indirectly* by means of technological machinery apparatuses which, as it were, intervenes between man and nature, and which is itself applied to already modified and prepared working materials. Only by historical reconstruction is it possible to exemplify a technology directed at untouched nature. The starting point for all subsequent stages was already determined by preceding technological actions. The present state of technological perfection, as evidenced by the concrete artifacts of our world, reflects this cumulative process.

In a general way, on the basis of the criteria of impact or transformation, one can distinguish between five types of technology: (a) in all production processes, the impact on the work object results in *material transformation*; (b) in various power stations an *energy transformation* occurs; (c) communication technology and data processing are characterized by *information transformation*; (d) in the case of traffic and transportation technology the desired effect is *change of place*; finally, (e) storage systems such as warehouses, accumulators, punched cards provide the function of *preservation* of materials, energy, or information (Ropohl *3.8*, 160–64).

(4) *Instruments, apparatuses,* and *machines* are the visible signs of modern technology. Through the instrumentalization of technological action man multiplies the strength of his muscles, the abilities of his hands, and the performance of his sensory organs. In this way he frees himself from the biological limitations of nature and creates a largely open field for technological action, the boundaries being only the laws of nature, limitation of resources, and destruction of the natural environment that can no longer be tolerated. It is remarkable that technological devices can be activated and controlled either by man or by other technological equipment. The functioning of a technological process is not necessarily an indication of how it was initiated. The degree to which man's physical and intellectual work is delegated to

machines or automata is usually taken as a measure of technological progress. The most perfect technological procedures are thus those in which the human element is minimal.

At this point, the ambivalent relationship between man and his technological creations is revealed: Man is, on the one hand, originator and ultimate decision-maker for all technological action. The most elaborate system can, after all, be turned off by a single push of the button. On the other hand, however, it is precisely the smooth running of technological processes that demands unconditional adaptation to it, and regards spontaneous behavior as only a hindrance. In decisive cases, therefore, one tries to exclude man as a potential cause of interference. To achieve the highest technological perfection, man must thus subject himself to the requirements of his own creations: "The lord of the World is becoming the slave of the Machine" (Spengler *1.28*, 90). Nevertheless, this fate is not simply a given fact. To be sure, preservation of physical existence always requires a certain amount of technology, however, the willingness to accept the inherent logic of perfected technological processes is an historically contingent phenomenon. This attitude made possible the systematic development of European technology and became an internalized norm during its expansion throughout the industrialized nations. Such a technology-oriented, impersonal and rationalistic view of life is by no means a matter of course. This is borne out by the difficulties which the developing countries, with their different cultural traditions, have in finding direct access to modern technology.

(5) The *changes of state* which result from technological intervention in the natural world are based on the fact that objects or processes are modified or newly created in a goal-oriented manner. In terms of the results obtained, one can further differentiate actions such as those which prevent, preserve, create, or eliminate certain conditions (Kotarbiński *3.6*, 23, 32). As modern technology illustrates, the actual face of the physical world can be radically

altered by such actions. In addition to the intended and desired changes of state, technological measures always cause unintended side-effects. Friction and heat loss is a typical example. So is the destruction of the natural environment, necessarily accompanying certain processes. Just as, from the biologist's point of view, there is no difference between weeds and useful plants, so, too, from the physicist's or engineering scientist's perspective, desired effects and undesired side-effects are not fundamentally different.

In this regard, any distinctions to be drawn involve value judgments. This becomes particularly clear in cases where what was initially regarded as an unwanted side-effect is later found to be useful. Conversely, it can happen that results initially intended may ultimately become side-effects. To be sure it is possible to distinguish between efficient and inefficient technological procedures on the basis of an input-output ratio. But what is decisive here is the non-engineering, economic and − in a broader sense − social assessment. Only by relating technological results to human goals and objectives is it possible to evaluate them as 'advantageous and useful' or 'disadvantageous and harmful'.

(6) The *application or utilization* of a technological system presupposes input and output conditions which are compatible with the biological and physiological abilities of man. However complicated a technological process may be in its internal details, its final result can be utilized for human purposes only at the level of sensory perception. Thus, for instance, the result of complex electronic switching processes in a computer can only be grasped in the form of numbers or graphic representations. Information obtained about micro-phenomena or cosmic systems by means of sensitive instruments also requires conversion to sensibly observable data before identification can occur. This is true, however, only in terms of the *total system*. The processes in various subsystems may very well remain hidden from our senses. Input and output variables of the various subsystems of a computer, for example,

are not usually perceivable without technological aids, since the processes utilized for the objectives of the total system occur below our threshold of perception. For their control one needs an understanding of the function of the specific subsystem along with appropriate equipment. Through phenomena such as these, which are hardly transparent to the average person, technology causes a certain alienation in terms of a world that is no longer self-evident and immediately understandable.

4. THE NEUTRALITY OF TECHNOLOGICAL MEANS

The efficacy of modern technology is based on the application of principles of the natural sciences, and on the insights of the engineering sciences. While queries and concept formation in the natural sciences aim from the outset at *theories* that are as general and precise as possible, the actual objective of the engineering sciences is the concrete *realization* of technological systems and processes. Despite this difference in objectives, both areas have in common the empirical method and the formulation of theories in mathematical terms. As far as basic research is concerned, it is therefore no longer possible to unequivocally distinguish between these 'two areas. Both strive, through an appropriate experimental arrangement, to investigate a given phenomenon as clearly as possible, and to describe it in appropriate mathematical terms. From a logical point of view, the formulations used have the character of conditional propositions: They state which effects (consequences) occur in the physical world, if certain causes (prerequisites) are met. If such rule-governed relations are known, then, by means of controlled interference, necessary causes can be created and thus desired effects obtained.

Natural interrelationships thus discovered can, by their very nature, serve arbitrary ends. The empirical and experimental approach, aimed at gaining scientific and engineering knowledge, is designed to precisely express process behavior. Such processes are quasi-blind to the manner of their utilization. Beyond their being subject to immanent laws, they offer no resistance against any application to which they may be applied. The amazing achievement of the modern empirical sciences has been the understanding of such laws to an extent heretofore undreamed of. Implicit in this achievement, however, is its potential for utilization in a negative way, and, in the case of armament technology, in a clearly destructive manner. In contrast to all previous times, the power handed to mankind by modern science and technology requires a degree of self-control which is conceivably beyond what man possesses; herein lies the real dilemma of our times.

Technological systems, produced by scientific and engineering knowledge, within the framework of natural laws can, like the aforementioned natural processes serve arbitrarily chosen purposes. Being material objects, technological artifacts are, in principle, available for any application. A nuclear bomb, for example, can be used as a weapon or employed in canal construction. Technology can, by means of exaggerated performance demands and machine-like work, degrade man to a mere cog in a man-machine system, or it can humanize the workplace or serve to protect the environment. The instrumental character of technological means, undefined as to content, renders them *methodologically neutral*, i.e., applicable for any chosen purposes.

This general statement, however, needs to be modified in three ways: (1) Technological processes and objects are certainly not always *factually* neutral. One will never, for example, be able to find a peaceful application of highly specialized military weapons. The range of applicability of a technological system is inversely proportional to its degree of specialization; the more specialized the design, the narrower its applicable latitude. While a hammer is a relatively versatile tool, modern technological devices and systems are, as a rule, tailored to a very specific purpose, which, as a result, they very effectively fulfill. Within the context of a given technological potential a decision for or against selection of a specific goal is possible. Once the choice of goal is made and the particular technological system is constructed, there is typically not much leeway as to its possible applications. As soon as a specific kind of technology comes into existence, it 'demands' appropriate utilization in order to recoup the capital investments involved. It is therefore understandable why, especially in the case of major technological projects, there is considerable inertia involved in their start up. It is difficult to avoid undesirable consequences when a road once taken is subsequently abandoned. Hence technological decisions once made develop their own 'internal momentum' and exert an influence on the future.

Technological action is by no means neutral when judged on the basis of actual occurrences and perceptible consequences. There are always certain side-effects which may frequently be ignored through attention to intended results, but which go beyond production and application goals. Technology, however, is no self-sufficient entity, separable from physical processes and from individual and social life. It does not originate from another dimension and is not a mere disposable means to desirable results only. While this has always been true, it has only recently, due to the ecological crisis and scarcities of materials and energy, become a part of the general awareness.

All technological action involves certain artifacts (instruments, devices, machines). For this purpose raw materials must be mined and transported, production devices procured, and the resulting product shaped. Yet these steps are only indirectly connected with the ultimate function which the final product may perform — the function of a transport system is to produce a change in location. They do, however, themselves result in a transformation of the material world. Additionally, operation of the final product exerts an environmental impact. The transport system, for example, may generate exhaust fumes. The more complex the technological system, the more support equipment required — roads, airports, repair facilities. It should be apparent from all of this that automobiles or airplanes, for example, are no mere neutral means, the only practical effect of which consists in providing convenient and fast transportation.

(2) Technology not only creates physical side-effects. Emotional and intellectual impacts are produced as well. It is thus absurd to consider technology as *psychologically neutral*. Because it is part and parcel of everyday life, technology necessarily influences thinking as well as feeling. During the stage of craft technology, this influence remained a part of existing cultural and social relations. With the Industrial Revolution the situation changed fundamentally; where every action aims at attaining maximum yield in minimum time with minimum expenditure, and where, thanks to technological aids and scientific methods, such a principle leads to even greater 'successes', the search for even more perfect means must finally become the norm. In this way, such efficient technological activity, with its internal rationality, runs the risk of becoming an end in itself, completely obliterating the original goals for which the effort was initially undertaken. In the extreme, goals no longer determine means, rather, existing technological means determine the ends to be realized.

(3) Finally, it would be a mistake to consider technological means as *socially neutral*. Because of his biological nature, man is dependent on a certain kind of technology to create a 'second nature' as the basis and substratum for his physical actions. Technological conditions cannot help but influence particular economic, social, political, and cultural conditions. Actually, there is practically no area of life in which one could not point out a direct or indirect influence of technology.

In the era of crafts technology, muscle-powered tools were still under the direct control of man. They could be used in a relatively

METHODOLOGICAL ANALYSIS

straightforward manner to attain 'externally' imposed technological goals. By contrast, the highly specialized and complex systems of modern technology constitute a closed realm with its own laws, producing the desired results, only if man is willing to yield to its own immanent logic. That realm, with its instruments, apparatuses, and machines, rather than being a neutral source of means, has separated itself from man and assumed the character of an independent force, shaping the face of our present times.

5. HYPOTHETICAL IMPERATIVES

The results of scientific research, as well as the insights of the engineering sciences, are concerned with law-governed relationships among physical world processes. While one type research is theoretical and the other practical, the experimental results obtained in both cases, usually expressed in mathematical form, have the same basic function: the indication, under stipulated conditions, of what follows what. In the absence of human involvement, knowledge of the applicable laws allows prediction of particular natural phenomena. Knowledge of the present configuration of the solar system, for example, allows one to predict a future eclipse. In the case of the engineering sciences, the creation of a set of initial conditions subsequently yields a technologically useful result. An example of this would be the ignition of a fuel mixture which, in an appropriately constructed combustion engine, drives the piston.

The natural and engineering sciences thus always provide conditional statements about functional, spatial and temporal relationships between particular empirically observable phenomena. Such statements of observed interdependencies between actual as well as hypothetical states are always *descriptive* in character. This basic descriptive character necessarily follows from the empirical approach itself, which investigates only interrelationships between actually occurring and objectively verifiable phenomena of the physical world.

Yet, there is a certain sense in which the engineering sciences also possess a *prescriptive* character. Whenever processes and phenomena are to be induced which would not naturally occur by themselves, the engineering sciences provide action prescriptions indicating how to proceed. This principle, in fact, holds true not just for the engineering sciences but for all the empirical sciences, where, as a general rule, results can be formulated in terms of conditional statements. For if it is known from which preconditions certain desired results follow, then one has only to establish those conditions to achieve the intended results.

Thus, it is always possible, if one has knowledge of the applicable laws, to apply measures appropriate to the envisioned goal state. From a methodological point of view, this is merely another version of the experimental method. Whereas, in the case of the natural sciences, the task is to determine which as yet unknown effects follow from known and given causes, with the engineering sciences one exploits interrelationships to obtain desired results indirectly, i.e., by creating corresponding causes. While the goal in the former case consists in uncovering, through appropriately designed experiments, new causal interrelationships, the latter case, by contrast, uses already known interrelationships for the creation of technologically useful results. Because of this difference between theoretical research and practical application, one employs only corroborated theories in the case of technological projects, whereas, with scientific research, one endeavors to examine particular predictions on the basis of yet unconfirmed heuristic hypotheses (Agassi *3.1* and Bunge *3.3*, 121–150).

The situation described here applies not only to technological subsystems, but also to such systems in their entirety, since their functioning, too, is always based on physical processes describable in terms of correlated conditional statements of universal validity.

METHODOLOGICAL ANALYSIS

By themselves, the engineering sciences thus provide mere *hypothetical imperatives*; they explain how, under well-defined conditions, to proceed toward the achievement of a certain goal. Yet, by their very nature they cannot provide any information as to *what* should be achieved with the existing resources. To make the step from theory to practice is to proceed from an understanding of the anticipated consequences of a particular procedure to a concrete directive for action. This step is rational only if, in addition to the theoretical knowledge, the intended goal is also stipulated.

This fact is of fundamental importance for the understanding of all technological activity. It implies that no circumstances whatsoever can by their mere existence necessitate a particular mode of action. Thus, in particular, one cannot, from the mere existence of a potential for technological action, infer that this possibility should be utilized. Only if man has, in addition to *descriptive* knowledge of existing or predictable physical phenomena, *also* a *normatively defined goal* reflecting a desirable change, does the necessity of a particular kind of technological action become apparent. When such is the case, one will indeed feel the "practical necessity" to do everything required to achieve the set goal (von Wright *3.10*, 67–68).

As von Wright further states, the *subjectively* made decision for a certain goal leads, on account of the objectively given circumstance and the technological action potential, to an action that "takes place of one's own will and is, at the same time, determined." The particular goal is the moving factor without which the action would not be initiated at all: concrete execution, then, is controlled in a specific manner by the information about the relevant (causal) interrelationships. This situation could be interpreted in such a way that one who wants to achieve a goal, must also want the means necessary for its realization (cf. 47–52). As applied specifically to technological action, however, three qualifications must be offered:

(1) One who aims at a certain goal must accept the means which lead to its realization only if there exists no other way of achieving the same result. Only then are we indeed dealing with necessity. From the standpoint of engineering science, however, there are usually alternative possibilities for solving a technological problem. In this case, therefore, the task is not to identify *one exclusively* appropriate procedure to be developed in detail and tailored to the given problem; rather, consideration must begin in a prior selection between the *different* applicable possibilities. In transportation or energy technologies, for example, various ways exist for mastering a certain task. For the engineer, the selection of optimal means precedes the actual construction phase.

Alternative solutions to a technological problem (alternative technology) may be approached in one of two ways. One the one hand, it is possible to retain the goals and employ alternative *means* to achieve them. On the other hand, it is possible to define completely new *goals* which may be realized with existing technological resources. An example of the first approach would be the case in which one retains the goal of individual transportation but replaces the automobile combustion engine with an electric motor. In the second case one could abandon the goal of individualized transport and adopt public transportation. The term 'alternative' applies only to the *manner* in which physical processes are utilized for human purposes. In terms of the *immanent structure* of any given material process, no alternatives exist, since such processes are governed by invariant and neutral physical laws in terms of which considerations of utility are irrelevant. There is, for example, no such thing as an environmentally benign physics, though one may speak of an environmentally careful technology.

(2) The situation can also arise in which an otherwise desirable goal may be abandoned if the *investment* cost involved is too high. Desire for a particular goal does not, after all, entail desire for

the pertinent means. Indeed, technological questions, as do situations in daily life, always involve considerations of cost or effort in the selection of goals.

As a rule, the tension between the desired and the actually attainable is reduced to a tolerable level by *adapting goals to what is feasible*. Yet it would be wrong to demand complete correspondence between desired goals and existent means, since this would imply that only that 'goal' could be envisaged for which all the means of its realization are already available. Indeed, throughout the course of technological development, projects have frequently been undertaken toward goals that seemed beyond reach, and it was only afterward, through appropriate research and development, that concrete means of attainment were worked out. Well-known examples include the planning of space programs or of new weapons systems. Admittedly, erroneous predictions have occurred. Nevertheless, such projects show that, with the systematic use of resources, technological development can, to a large extent, be planned.

(3) Finally, one must allow for the possibility that while the immediate results of a particular technological action may indeed be both achievable and desirable, its *unavoidable side-effects or secondary consequences* are such that it must be abandoned. This is particularly the case when technological processes exert undesirable effects on the environment, as in cases of damaging noise levels or air and water pollution caused by the discharge of harmful substances. Here, however, by some means of comparison, the decision must be made as to whether the advantages gained outweigh the negative and harmful disadvantages. As modern technological procedures continue to increase in size and intensity, their intervention into the well-established equilibrium of natural processes and their creation of correlated side-effects continue to grow. Self-restraint in the application of technological means, thus, assumes an ever increasing importance.

A subjectively compelling need to act requires both knowledge and a goal. This statement is self-evident, in need of no further elaboration. In the area of technology, as in other areas of life, mere *knowledge* of the consequences resulting from a certain action is insufficient to establish the necessity of that action. Yet, in the case of modern, highly specialized technology, the fact that a multitude of far-reaching measures already exists, in forms such as production and transportation systems or supply networks, serves to narrow down the actual range of options, thereby more or less dictating a specific manner of action. In addition, certain goals, such as relief from physical work, increased productivity, or increasingly efficient technological systems, are considered so self-evident that they are no longer explicitly articulated or questioned.

In this situation *reflection on the goals* that are explicitly or tacitly pursued in connection with any particular technology can serve an *enlightenment function*: it can clarify which existing facts one accepts as simply given and unalterable when, in fact, their (gradual) modification is possible; furthermore, through explicit formulation and discussion of goals, it can reveal intentions which are otherwise only intuitively or implicitly assumed in the pursuit of specific technological actions.

The familiar claim that existing technological means are what determine ends to be realized can now be reduced to its essential core. The fact that a web of rational and purposeful action runs through all technology is not refuted by deviating practices. There are instances where indeed the available resources and the technical potential resulting from research and development are used in an unreflective manner, just because they are there. Nevertheless, even in these cases there are hidden goals which serve as a guide to the action.

6. TECHNOLOGICAL PROGRESS

Whereas wishes and needs can multiply practically without limits, the possibilities for satisfying them are always limited. The scarcity of resources therefore necessitates prudent and economical management: means that have been used up on a certain task are no longer available for other uses. In order to be able to fulfill as many needs as possible, one therefore strives to achieve each particular goal with a minimum of effort. The "rationality principle" of technology (von Gottl–Ottlilienfeld *2.1*, 211) dictates accordingly that that road always be chosen which yields maximum output for minimum input.

This principle provides the basis for determining in each instance what constitutes a 'better' technological procedure: we are dealing with technological progress when either the same work can be done with less input, or when one achieves higher output for the same input. It is in this sense that the efficiency of a technological system is frequently taken as an indicator of technological progress. Specific forms which this progress may assume include the construction of completely new systems, or the improvement of existing systems in terms of increased operating life, reliability, sensitivity, precision, or speed of operation as well as faster and/or cheaper production.

All technological innovations are based on the creation and appropriate combination of technological processes in such a way as to yield the desired utility. Natural processes can, within the limits of their immanent laws, be utilized to achieve arbitrarily chosen technological objectives. At first glance any particular innovation constitutes nothing more than a mere alternation. For a new procedure to be judged as a technological progress, very specific *evaluation criteria* are also required. Only then is it possible to determine whether a particular innovation constitutes a step forward or backward.

There are many instances of invention and innovation which embody a novel technological principle, but, because their practical application is not considered desirable, are nevertheless not utilized. Yet, leaving aside considerations of direct, practical utility, one may still speak of progress insofar as the state of technological knowledge and ability has been increased. To speak of technological progress, i.e., a positively assessed instance of technological change, it is necessary that at least the following three distinctions be made:

(1) A successful increase in the range of technological knowledge and skills constitutes progress in *engineering science*. The decisive criterion is the expansion of the range of possible action by means of which natural processes may be utilized for human purposes: every innovation or invention contributes in principle to an increased repertoire of potentially useful technology, regardless of its immediate practical utility. Progress in the engineering sciences is thus a necessary, but by no means sufficient, prerequisite to realization of the corresponding technological procedures and systems. In terms of the level of technological know-how, it supplies a *basis* for the construction of artifacts in terms of increased yield or efficiency.

(2) *Actual* technological progress is achieved when a certain part of advances made in the engineering sciences is realized by means of largely economic mechanisms. In more recent times there has emerged an unmistakable trend toward realizing everything that the present state of knowledge and ability will allow. Unlike the engineering sciences, the norm here is not merely one of an increased domination of nature, but rather one of more *efficient and economical* utilization of work and materials in applying the pertinent physical processes to human purposes. For example, a procedure, which, from the engineering science point of view, is quite advantageous in terms of energy yield, may nevertheless be dropped on economic grounds – raw materials or construction costs may be too high. In this case, in contrast to the engineering science perspective which is only concerned with physical processes and technological procedures, a cost/effect

assessment either implicitly or explicitly occurs in which cost is measured against yield. Questions of economics, however, are always a part of a broader selection process, which is the ultimate determinant of the desirability and 'progressive character' of any envisaged technological innovation.

(3) The uneasiness frequently found today in connection with technology suggests that actual changes do not necessarily always lead to improvement. Besides actual technological developments, one must further consider *ideally* desirable changes, which would constitute technological progress in the proper, positive sense. Regardless of what the specific embodiment of this normative concept of progress might look like, it can only be based on a comparison between the desired later state and the preceding earlier state. Here, as was the case with (2) above, careful consideration of all relevant factors is necessary. Any critique of actual practice involves an examination of the *criteria* considered relevant at the time — criteria which are essentially based on short-term goals, quantitative results, and the physical results of technological actions. Long-term cultural and social effects as well as impacts on the earth's life support capacity are rarely if ever considered. A genuine appreciation of these problems, however, is only to be gained if one examines the historical process that led to the present situation.

CHAPTER IV

THE ROAD TO MODERN TECHNOLOGY

1. THE SOCIO-CULTURAL APPROACH

The scientific orientation of the engineer and the historian of technology leads them to regard technological processes and systems as isolable phenomena, capable of being analyzed strictly in terms of the principles of their constructions and functional mechanisms. The human actors through whose agency these artifacts are created, the motives which drive them, and particularly the effects of technology on social, cultural, and intellectual life are, for the most part, disregarded. This approach, which is both abstract and, in an objective sense, also concrete, is fruitful and even indispensable in clarifying the specifically technological side and the methodological principles of the technological action.

However important such an analysis may be for investigating immanent technological questions, it nevertheless does not touch on the real problems of technology. This is particularly evident in the impersonal style of the natural and engineering sciences. Here technological phenomena are treated as objectively given and linguistically characterized in the third person, though, in fact, they only come into being by means of human action (Pierre Ducassé *1.12*, 27). The determination of goals in particular, to which technological measures are the response, along with the repercussions those measures have on man, are treated as external variables, supposedly 'outside' the objective process of technology. In reality, however, the complex phenomenon which is technology is always *simultaneously* a utilization of natural forces and a socio-cultural process. Either a 'technocratic' or 'sociocentric' approach is possible, depending on where the accent is placed. On the whole, the major weight should clearly lie on the side of social and cultural conditions and technological consequences, since it is on this basis that engineering activity is ultimately assessed.

When one focuses more closely on the interaction of man and technology, it becomes evident that man is at once both creator and creature of technology. There exists a reciprocal relationship, which, on an abstract level, can be broken down into two components: (1) the generation of technology by man and (2) the forming of man by technology (which he himself has generated). A balanced judgment requires that *both* components be considered. They are in fact only separable in concept, not in essence. On the one hand, man as the *"as yet undetermined animal"* (Nietzsche 4.22b, 74), by means of technological activity and artifact, creates his concrete environment in ever new and historically changing forms. On the other hand the execution of this technological activity along with life in the 'second nature' thus created exerts its own reactive force on man. One of the crucial questions concerning this process is how its effect can be distinguished from other social, cultural, and intellectual factors.

2. HISTORICAL DETERMINATION

The physical processes that occur in technological systems, and the methodological principles underlying their generation and utilization for human purposes, are defined by the structure of the material world. The immutability of natural laws lend a 'timeless' character to physical processes. Despite the ever-changing status of *subjective* knowledge, the *objective* structure of physical processes is transhistorical. For social and cultural processes, embedded in the concrete historical situation, no comparable invariant laws can be stated. While physical events may be studied in isolation, by means of repeatable experiments, history consists of unique events which are always embedded in a more comprehensive context.

The odds against reaching universal statements for the *historical* development of technology, and thereby transcending summary description or narrative, are by nature greater than in the case of methodological investigations. Yet, even here a structural analysis is indispensable, if we wish to conceptualize the past. The past does not, after all, speak to us by itself. Only by means of a definite system of concepts, appropriate criteria for distinguishing important and unimportant events, and a general theoretical framework, can historical events be transformed from an arbitrary conglomeration of data into a meaningful whole. While different theories will result in a variety of presentations and interpretations of technological development, (Hübner *1.15*, 14), the object of investigation provides a common reference point, in terms of which the adequacy of particular theories may be measured as to their content.

History records a great diversity of advanced cultures, each of which had perfected the arts and crafts to an amazing degree. Industrial technology, which today determines the fate of all mankind, arose, however, during a narrowly circumscribed phase of European history and spread from there across the entire world. This raises the question as to whether the specific constellations of Western history necessarily and inescapably led to technology in its modern form. Such a quasi-teleological interpretation, in which all the relevant previous events are interpreted as preparatory steps in a process leading necessarily to technological development, can *ex post* always be constructed. Its truth, however, is empirically unverifiable.

This becomes clear if, instead of the nexus between past events, one visualizes the connection between the present situation and

THE ROAD TO MODERN TECHNOLOGY 69

future states. It is by no means clear, for example, what the 'necessary' consequences of the present-day state of technology will be. At the moment we can observe both strong expansionistic and counter trends. Another fact also seems to weigh against the quasi-teleological position. During the Middle Ages, the technological development of China and the Arabic countries was quite comparable to that of the European countries. Yet it could not be predicted with certainty that European technology alone would lead finally to the Industrial Revolution. The indeterminateness of future development, as illustrated by these examples, is not a reflection of an insufficient level of subjective knowledge, but rather an illustration of the objective openness of the historical process itself.

Even though it is impossible in historical investigations to state the *consequences* to which a given event must necessarily lead, it is nevertheless possible to show the *conditions* necessary for a certain historical development. It is possible, for example, based on the fact that between 1740 and 1840 the Industrial Revolution took place in Europe, to infer the indispensable ideological, physical, and social preconditions without which this event in its actual form would not have been possible (cf. von Wright *3.10*, 61–70).

The preconditions which finally *made possible* industrial technology can thus be identified by a retrospective look at the actual course of events. One cannot, however, prove that these preconditions necessarily had to lead to this particular result. Keeping this qualification in mind, we shall investigate, later in this chapter, the socio-economic, technological, and intellectual prerequisites for the development of modern technology. The lack of a strict historical determinism does not, however, preclude the fact that within historical situations more or less definite temporal trends and tendencies are at work which will, in all likelihood, lead in the forseeable future to fairly predictable results. Such trends, however, do not prescribe a necessary course to history, and therefore do not provide the possibility for infallible predictions either.

The developments which finally lead to industrial technology, are best viewed as mere aspects of a unified whole. Any "one factor theory" which focuses on only one of those aspects is doomed to fail (cf. Lenk-Ropohl *1.20*, 112). Yet methodological considerations force us to introduce terminological differentiations and to divide this complex issue into coherent subcategories. Only this will allow us to gain insight, through systematic analysis, into the otherwise vague notion of a 'unified whole'.

3. MAGICAL AND TECHNOLOGICAL THINKING

The fact that modern times are referred to as the 'technological age' underscores the crucial importance of technology today. This designation, however, can easily obscure the fact that technological activity has been going on as long as mankind has existed. In order to safeguard his existence, man has always relied on appropriate procedures and tools. The tasks involved have not fundamentally changed throughout history. True, they have become more varied as new possibilities have given rise to new needs, which are, in turn, satisfied today by means of highly complex, efficient (though correspondingly trouble-prone) devices. Yet, man's basic biological needs remain invariant. Man's survival in a largely hostile environment requires food, clothing, shelter, mobility, communication with others, defense against one's enemies, and the opportunity to recuperate from illness. In this regard, the introduction of the bow and arrow for hunting or as a weapon was as important as was the invention of gunpowder; discovery of the wheel or the breeding of horses was every bit as momentous as the invention of the automobile or airplane. And was the building of mud huts any less important an innovation than the construction of a skyscraper?

From the beginning, all technological procedures have been characterized by their *distancing* from the concrete situation. Unlike the animals, which always turn to the immediately given possibilities of their environment, man is not satisfied with what is at hand. He steps back, as it were, intent on some creative transformation. Goals, rather than being directly achieved, are approached circuitously. Immediate satisfaction of needs gives way to technological measures that can be complex and protracted (providing working material producing tools and subsidiary equipment) and will only at the very end produce the desired product. Through technology man "negates" nature as he finds it by consciously and purposefully transforming it according to his ideas and wishes into a "supernature" (Ortega y Gassett *1.25*, 25–26).

Magical procedures constitute technology in seminal form. Consider, for example, the prehistoric role of metallurgy. The central figure is the smith who alone knows the secrets of the metal. The practice of his craft, so crucial for subsequent development, is accompanied by magical practices, rituals, and generally ambivalent symbolism, reflecting an awareness that the production and processing of iron amounts to human action taking a new and dangerous direction, a sacrilegious interference in the divine order of nature, yet, at the same, time the promise of a higher stage of culture (Mircea Eliade *4.8*, 24–33). The 'wisdom of myth' always understood the ambivalence of technology, a fact only too obvious in our own highly 'technologized' world. Thus, Hephaestus, the god of the smiths, is admired for his ability, but his activity is considered, at the same time, also cunning and ominous. The punishment for technological *hubris* is symbolically expressed in the mythical figures of Icarus and Prometheus.

Ellul is another who has noted the structural similarity between magic and technology. Both activities, for example, attempt to achieve a given goal as reliably and simply as possible. Just as one tended to blame the magician rather than the procedure when, say, an attempt at rainmaking failed, so too, Ellul proposes, modern man's admiration of technology can be explained as a deeply rooted awe felt toward the amazing and mysterious processes designed by his own hands (*5.4*, 24–27).

Yet, superficial similarities notwithstanding, magic and technology are different in several crucial ways. As Cassirer has demonstrated, there are basic differences in the actions of man, in his relationship to nature, and in the range of possibilities for his activities. In magical practices the desired event is mentally anticipated in ideal form, and represented through magical words and images. The more accurately such anticipation takes place, the more likely, according to this understanding, the success of the action contemplated (campaign in war, hunting or fishing). Actual events cannot, after all, escape this concentrated spell. However, while magical thinking has already come to understand a certain order of natural processes, they are conceived in totally anthropomorphic and animistic terms. Yet, a separation from the immediate setting has occurred; by creating on his own an image of a desired future event, man has taken his first steps toward a conscious and planned shaping of the world.

Still lacking in this approach, however, is the objective and sober distancing from natural processes which characterize the modern scientific and technological world view. The magical identification of the Ego and the world forestalls the consideration of natural processes as existing in their own right, independent from man and subject to their own specific laws. It obscures the fact that such laws themselves limit the possibilities for human control. In fundamental contrast to the magical conception of nature, modern science and technology explicitly investigates the lawlike structure of natural processes, thus making possible the systematic utilization of natural forces (Cassirer *1.9*, 31–34). The technological-scientific view of nature is, thus, at once more modest and more presumptuous than the magical point of view: on the one hand, it accepts in advance, as magic does not, the fact that the possibilities for human influence are limited. On the other hand, however, with an attitude quite alien to magical and animistic thought, it completely exploits and comprehensively transforms nature *within* such limitations, which are furthermore pushed more and more outward through systematic research.

Cassirer rejects as subjective and unscientific the magical world view, in which "man's own wishes and speculations are projected" into nature (*1.9*, 34). As a matter of fact, no one today is likely to attempt to solve scientific and technological problems by magic. Yet this irreversible process of enlightenment has had its price. No longer do we consider nature as an animate and sacred cosmos into which man is integrated. Nature is confronted, rather, as a neutral conglomeration of matter opposite us which can, within the limits of physical laws, be treated as we wish. The resulting technology, despite its undeniable positive achievements, has created the possibilities of a nuclear apocalypse and a devastated natural environment. Are we prepared to claim, at this point, that man's radical emancipation from nature constitutes unqualified progress?

Hübner has observed that mythological and scientific experience are incommensurate. Since each is based on an all-embracing a priori model of the world, there is no superior criterion of truth to which we may appeal (*4.13*). It is, however, an established fact that the rational and scientific approach has been pursued for thousands of years in Europe, and has subsequently become the norm for all of mankind. *Outwardly* the superiority of the scientific world view and of its impressive technological achievements is undeniable. Because man has been spending all his energies on mastering and using nature, however, his *inner* problems have necessarily received less attention. This area is thus, on the whole rather impoverished, in obvious contrast to the fields of science and technology.

Present-day threats to the ecological balance of nature and to the presence of an adequate supply of mineral raw materials and fossil energy have prompted appeals for a return to a 'brotherly' and sacred view of nature. Such a view, which was, for example, characteristic of Francis of Assisi and was still echoed in the Romantic Era would, however, require a fundamental change of modern rational and enlightened consciousness. In fact, the process of turning nature into a mere impersonal object cannot easily be reversed. Because man's relationship to nature is no longer determined by a spontaneous affective experience, but is based instead on a rational calculus, technological self-restraint, that has become inescapable, must be guided by conscious reflection instead of immediate experience. In other words, the only viable option appears to be to commit oneself totally to a rational approach and, in precisely this manner, to advance toward a more appropriate view of nature. Heinrich von Kleist* expressed this thought when he wrote that we have "to eat of the tree of knowledge again to fall back into the state of innocence." This he considers the "last chapter in world history" (*5.19*, 831).

4. SOCIO-ECONOMIC CONDITIONS

The character of modern technology can be clarified if one examines its historical development up to the present. Such an analysis can also facilitate a more appropriate assessment of the possible ways to influence future technology. On the face of it, however, any such historical review can only provide information about the *temporal coincidence* of certain events. Additional claims about *causal dependencies* need to be based on the knowledge of pertinent cause-effect mechanisms; however, this, as a rule, instead of

---

* Early 19th century German author, best known for his dramas and tales (Translators' Note).

THE ROAD TO MODERN TECHNOLOGY

being empirically verifiable, can only be hypothetically stipulated. As far as the development of technology is concerned, problems of this kind become especially crucial when one attempts to clarify the relative importance of ideal and real factors for the rise of modern technology.

Regardless of how the question of the 'ultimate' causes of technology is answered, certain necessary social and economic *preconditions* for a specific technological activity can be identified. Any technological activity which goes beyond elementary and individual measures is, by its very nature, of a *social* kind. Its execution and its effects concern not only the individual but also entire groups of individuals, and in the limiting case society as a whole. Production, distribution, and consumption of technological goods is mediated by economic exchange processes and executed by a variety of actors. Any increase in efficiency, and thereby also in system complexity, depends on a corresponding high degree of division of labor. Work processes become increasingly subdivided, involving additional numbers of individuals needed for rather specialized tasks. As a result, as technological developments intensify, the interweaving of the technological and social process in both the work sphere and in private life, becomes closer and closer.

In this phenomenon, a double dependency can be observed; on the one hand, technological innovations can cause social change. Well-known examples are the formation, in the early days of industrialization, of the industrial proletariat, or the serialization and rationalization of production processes which have made possible a uniform mass civilization. On the other hand, technological developments are themselves dependent on existing social conditions, since in any specific era technological innovations which according to knowledge and ability may in principle be possible, will only be realized to the extent that the requisite social and economic structures exist or can be created in support of technological 'necessities'. This reciprocal relationship, essentially shaped by the particular technology involved, may be referred to as 'socio-technical system'.

Various causes for these historical changes have been proposed. Marxist theoreticians see the actual dynamic force as the process of production of material goods which forces social processes to change accordingly. The heuristic value of this view is beyond dispute. In the nineteenth century Marx perceptively recognized both the significance of technology for modern economic and social development and the dynamics of the capitalist mode of production. His doctrine of the four historical stages of societal development — slave holding, feudal, capitalist, and communist — is, however untenable, as is the exclusive interpretation of social history as one of class struggles. These theses have accordingly been variously weakened or even discarded by Neo-Marxists (cf. Helmut Fleischer *4.9*). With the orthodox Marxism of Eastern Bloc countries, however, the development of technology, which will bring with it material abundance, has become the decisive point of reference and the government-propagated norm. In Lenin's succinct formulations, "Communism is Soviet power plus the electrification of the whole country" (*5.24*, 419).

Werner Sombart holds that capitalism is the creation of outstanding entrepreneurs rather than the result of a "'collectivistic' or, as it were, organic process" (*4.27*, 836). While the village community or the guilds owe their existence to such organic and anonymous processes of growth, capitalism is based on rational, well-calculated, and far-sighted planning, which is always the creation of the entrepreneurial personality. The revolutionary innovations of these bold men represent a complete break with the traditions of peasant economics or of the vocational and commercial artisans: what becomes most important is calculation aimed at expansion and profits rather than the "principle of satisfying needs" (836–837). The interpretations of Marx and

Sombart are not, however, necessarily contradictory. They may be reconciled if conceived of as abstract and collective versus concrete and individual explanations of socio-economic change, respectively.

Sombart's theory of the dynamic drive of the capitalist system which is always bent on continuous change and expansion is keyed to the entrepreneur, a person both economically and technologically creative. Werner von Siemens would be an example. In the Western industrial nations today, however, the free-wheeling entrepreneur has largely been replaced by managers, trained to perform any one of a number of functions, and, interestingly enough, little different from the functionaries of the planned economies, with their preoccupation with routine and anonymous office and committee work. Thus, at present, economic dynamics in both private capitalist market economies and in state capitalist planned economies largely reflect institutionalized mechanisms of scientific and technological growth rather than the spur of active and creative individuals.

In the future, according to Schumpeter, the role of the entrepreneur will increasingly lose its importance as "innovation itself is being reduced to routine," and accurate economic calculations replace the "romanticism of earlier economic adventures" and "ingenious intuition." Furthermore, there is today no longer a need to overcome producer and consumer resistance against new products. The sheer fact of their newness is accepted by many as prima facie evidence that they are improvements, even if, in terms of reliability or durability, they are actually steps backward (*5.35*, 132–133).

Bert F. Hoselitz, using the development of Japan and the Maori in New Zealand, has shown that economic change need not depend solely on the interaction of class structure and mode of production nor on the "social deviation" of active entrepreneurial personalities. It can also be achieved by means of a new social elite.

In this case, the traditional social structures are retained essentially unchanged, but are reinterpreted by the new leadership stratum in the sense of altered value concepts. Hoselitz sees in this kind of economic change the possibility whereby developing countries can reach a satisfactory economic and technological level without having to imitate social conditions of the industrial nations (*4.12*a, 74—84).

The origins of rationalized methods of economic management, which are characteristic of industrial technology, can be traced to the medieval cities of Europe. With their merchants, guild artisans, and cooperatively organized *constituency of citizens*, with its sharp legal separation from the surrounding *agrarian society*, these communities provided a fundamental contrast to other cultures in which such a separation is unknown. Through struggles between Church and State during the height of the Middle Ages, the worldly and religious spheres also became separate. This differentiation and secularization laid the groundwork for the ideas of the sovereign state and the modern representational political system. It made possible, as well, the unfolding of a work ethic oriented purely to this world, which in turn decisively fostered economic and technological development in a capitalistic economic mode. Thus, medieval capitalism of trade and finance, with its penchant for exact planning and calculation, can be considered the forerunner of later industrial capitalism (Otto Brunner *4.4*, 91—101).

The individual nations of Europe did not all participate to the same degree in the events which were in time to lead to the unfolding of technology: the voyages of discovery that led to the building of colonial empires, to overseas trade, and thus to capital accumulations, were the achievements of Spain and Portugal. The introduction of new forms of energy and of mechanized production began in England, to be subsequently adopted by the other European countries, one by one. At about the same time, with the French Revolution, the ideal of representative democracy emerged openly into the political arena and, in time, became generally recognized as the political norm.

It is no exaggeration to say that the present time is essentially determined by two principles: one, the idea of *democracy* with its ideal of equality and the sovereignty of the people, and, two, the concept of modern *technology*, which originated in the combination of the experimental mathematical sciences and industrial production. When these principles conflict, it is usually the technological and economic 'logic of things' that is given priority over the basic rights of individual freedom. The connection between industrial technology and egalitarianism is evident: both originated from the same intellectual source, namely the objective and world-immanent rationalism which culminated in the Enlightenment. From a practical point of view, the leveling of incomes and taste forms the prerequisite for mass production, without which efficient industrial technology would be inconceivable. Modern technology, in turn, creates the possibility of realizing, to a large extent, the ideal of equality in material terms.

As noted by Schumpeter, the actual stimulus for the capitalist economy can be seen in the dynamic "process of creative destruction" through which old structures are continually destroyed and new ones built. This involves new consumer goods, new methods of production and transportation, new markets, or new forms of industrial organization (*5.35*, 81–86). While this ongoing process of development and renewal is caused in the Western industrial nations by competition inherent in the market mechanism, and by the profit motive, this is not the case with the communist system of planned economy. Nevertheless, Schumpeter's description of the dynamics of industrial technology applies in its basic outline to these countries as well. It thus becomes clear that modern industrial technology, set in motion by capitalistic economic forms, has developed a 'momentum of its own' on a world-wide scale which is largely independent of the economic system and which cannot be described with economic categories alone.

5. TECHNOLOGICAL FOUNDATIONS

From the perspective of man as acting *subject*, technological action is always tied to particular social structures and economic processes. In terms of the physical world as object to be worked on, those prerequisites find their counterpart in the particular state of technological resources, since utilization of natural resources is possible only to the extent that the necessary material means (raw materials, tools, energy) and the requisite intellectual knowledge (artisan rules of experience and/or science) are present.

Although gear constructions were known to antiquity, mechanical technology was, nevertheless, hardly developed. The mechanic Heron of Alexandria, who lived during the first century A.D., built many playful devices which had no practical significance beyond their use for cultic purposes, e.g., as holy water dispensers or temple door openers. Such mechanisms did not require significance until the Renaissance and Baroque Age, when they were used in the construction of elaborate mechanical clocks, instruments, and animated fountain figures of princely gardens (Friedrich Klemm *4.16*, 35–42). As early as the first century B.C., waterwheels replaced human muscle power for the grinding of grain. Yet it was not until more than one thousand years later that water-driven saw mills, rolling mills and trip hammers were used. In the thirteenth century windmills could be found throughout Europe. By the late Middle Ages the basic principles of rocketry as well as of gunpower were known. The first tentative steps toward utilization of compressed air or steam as a power source occurred in the Renaissance, and these can be viewed as the forerunner of the steam engine of the 17th century (Lynn White Jr. *4.30*, 79–103).

The crank, an indispensible device for the utilization of rotary motion in machine technology, is first observed in Europe around 800 A.D., used with grindstones. Not until six centuries later is its use broadened to the spanning of crossbows and for winding skeins of yarn. Very soon it appears in the form of the breast drill, offset axle, and in combination with piston rods and flywheels (*4.30*, 110–115).

It is not until the end of the Middle Ages that the crank is finally utilized in a wide variety of applications. The surprising slowness of its diffusion into general practice is explained, White believes, by the fact that it involves a transition from organic, human motion into inorganic, mechanical movement. That is, since the rotary motion of mechanical devices is unknown in nature and is seemingly in conflict with the natural reciprocating motion of normal body movement* its potential utility has only slowly gained general acceptance (114–115).

This example clearly illustrates the absurdity of trying to explain the historical development of technology by merely projecting present-day attitudes, conditioned as they are by current innovations and mechanical processes, backward through time. The manner in which the given technological potential is utilized in any particular era does not depend upon unchanging anthropological factors which require the realization of all possible actions. Implementation, instead, reflects the ideas about human nature and the world held by each particular culture. In some cases those ideas appear so foreign to us, and that culture's failure to appreciate what seems so obvious an improvement, appears so puzzling to us, that we find it virtually impossible to identify with them. How else are we to explain how apparently simple, and from our vantage point, obvious technological innovations such as

---

* White notes (p. 115) that the great physicist and philosopher Ernst Mach noticed that infants find crank motion hard to learn.

the wheel, the stirrup, power mills, or the crank took so long to develop and spread to general use. As an example, Alexander Rüstow mentions two-wheel ox carts with solid disk wheels rigidly affixed to an axle turning underneath the cart bed. Such carts have remained unchanged in form for more than five thousand years in what is today Western Turkey (*4.25*, 58–59).

Mechanical inventions of the late Middle Ages also include the pendulum (as a regulator of periodic motion), the pedal (for driving looms, lathes, and spinning-wheels), the connecting rod, flywheel, and power transmission by belt drive. Especially remarkable, however, are weight-driven, and in smaller versions spring-driven, mechanical clocks. Their complex gear trains, elaborate representations of the heavenly bodies, and richly decorated figurines, appealed to the spirit of the times, with its fascination for things mechanical. It is thus not surprising that these clocks created a sensation. By the end of the fifteenth century, White states, there existed in Europe more machines for the generation of power, along with more knowledge of energy production, transmission, and utilization, than ever before. He even suggests that, in terms of fundamental innovations, more was accomplished in the Middle Ages, than in the four centuries after Leonardo da Vinci; White further claims that up until the discovery of electricity all the basic principles in use originated during the four centuries preceding Leonardo (*4.30*, 129). Whether or not one is prepared to agree that first tentative beginnings have greater significance than mature and practical constructions, the examples we have described demonstrate the large extent to which modern mechanical technology is derived from its medieval forerunners.

Interest in technological discoveries in the sixteenth and seventeenth centuries is further reflected in the literature of the times and in many proposals for new inventions. Yet, the fact is, their view of technology and ours are different in one fundamental way: nature was for them an animate organism, inhabited by

THE ROAD TO MODERN TECHNOLOGY 83

demons and mysterious magical forces. The act of invention was itself a mysterious process, one based on divine inspiration, not sober empirical investigation and related technical considerations. Accordingly, as late as the seventeenth and eighteenth centuries the majority of technological inventions — some with an air of fantasy about them — were more the work of amateurs than of experts. The fact that invention was possible without detailed technical knowledge illustrates how close technology and everyday experience actually were. By contrast, today's technology, based as it is on science, has become entirely the domain of experts.

It is true that the era of empirical science, with its utilization of mathematics and controlled experiment, may be traced as far back as Galileo. Yet, the traditional animistic view of nature, as expressed in magical thinking, alchemy, and the witch trials, persisted, giving way only gradually to the emerging empiricist and rationalist ideas. In fact, mathematics and astronomy, which were the first sciences to emerge out of the struggle with natural philosophy, initially exerted no direct or practical influence on technology at all. Technology continued in the artisan tradition. Themes which predominated in the technological literature of the day include alchemy, mining, navigation, along with pyrotechnics and artillery (Sombart *4.27*, 463–75).

Artisan technology is based on traditional skills and rules of experience which have been developed over a long time and which continue to prove effective. The relevant question is: *How* is one to proceed in order to achieve the desired goal? Each particular procedure is considered only as means to a given end, and not as an object of theoretical scrutiny itself. In contrast, scientific thinking is interested in *why* a certain measure has precisely this and no other effect. By its very nature, the tradition-based practical orientation of technological activity is not conducive to fast or fundamental innovations, since there is a subjective lack of psychological interest, combined with an objective lack of the necessary methodological prerequisites.

To systematically improve on established procedure, one would need to methodically analyze and vary the relevant variables involved. Thus, before the advent of the modern empirical approach, which questions every practice and neither practically nor theoretically acknowledges traditional authority, the intellectual prerequisites for industrial production and rapid technological change simply did not exist.

External factors as well contributed to the conservative character of artisan technology. Each craftsman obtained his materials largely from his own rural area, and, given the firmly established stratification of society according to estates, could rely on a fairly constant pattern of demand. Appropriate organizations (guilds and craft fraternities) strictly regulated the practice of crafts, membership quotas, markets and supply sources. The system of artisan technology thus lacked direct and basic competition, which is the economic stimulant of the capitalist mode of production (Sombart *4.27*, 204–11).

6. THE INDUSTRIAL REVOLUTION

Although the Industrial Revolution began in *England*, spreading then to other European countries, it was *France* that clearly had the lead in the theory of the engineering sciences. Nearly all theoretical work on power engines, civil engineering, strength of materials, hydrodynamics, and construction procedures originated with the French, but failed at first to influence technological practice (Peter Mathias *4.18*, 66). The advances in technology which made the Industrial Revolution possible were not the result of theoretical considerations and *deductive* derivations from general principles, but rather the result of practical attempts and *inductive* generalizations. One of the few exceptions to this

THE ROAD TO MODERN TECHNOLOGY 85

rule involves the application of the mathematical theory of the pendulum, developed by Huygens, to time pieces (Maurice Daumas 4.6, 4). Thus, the early achievements of industrial technology combined artisan skill with systematic experimentation, but were always aimed at immediate practical application. For the time being, scientific investigations, which were predominantly in mechanics and astronomy, exerted no *direct* influence on the development of technology. Indirectly, however, their impact was twofold.

In the first place the need for *exact measurement* in mechanics and astronomy placed increasing demands on instrument makers and clock builders who successfully combined exquisite artisan skill with mathematical knowledge. As a result, practical tasks such as navigation, surveying, and astronomical prediction benefitted from the availability of precise instruments. A. Rupert Hall observes that we owe these instrument makers the fundamental insight that a machine is able to work more precisely than the human hand (*4.10*, 338).

Secondly, technological development was indirectly influenced by the scientific orientation of the *intellectual climate* of the times, with its emphasis on empiricism and practical utility. By means of systematic collection, classification, description, and publication, the prerequisites for mathematical theorizing were established. The spirit of the times stressed the design of experiments and critical analysis of theoretical constructs; a large number of academies and scientific societies were newly founded to aid in the solution of practical tasks and to improve technological procedures. While technological innovations were not the result of direct application of scientific theory, neither were they based on random or unreflective efforts. The principle of trial and error, which led during the Industrial Revolution to workable machines and motors, reflected, despite its practical orientation, the indirect influence of the scientific mind-set.

The first phase of industrialization rather than being characterized by spontaneous new inventions or direct applications of scientific research, reflected systematic and detailed improvements on already existing procedures (Daumas *4.6*, 9). In the ensuing process of industrialization, the time lag between conception and application of technological improvement is progressively shortened. Not before the second half of the nineteenth century, however, could technological practice be characterized as applied science, although with subsequent developments this would increasingly come to be the case.

On the whole, the Industrial Revolution is characterized by its refinement of existing procedures or their large-scale utilization. In retrospect, the transition from manual artisan work patterns to the large mechanized factory with its division of labor, can be viewed in terms of technology as part of a relatively continuous development from earlier stages. It has proceeded at an accelerated pace into present times. Division of tasks, for example, was already present in pre-industrial manufacturing, as was the replacement of human muscle power by power mill energy derived from wind and water. In fact, the pervasive and varied use of mills in the period just preceding industrialization has prompted Sombart to characterize this era as the "classical age of the mill" (*4.27*, 485).

Thus, in both theoretical conception and practical application, industrial technology does not mark a radically new beginning, but rather a step along the road prepared and prompted by preceding developments. Yet, this step, which sharply increased the momentum of technological development, has proven to be decisive for the whole of mankind today. Therefore, it would be more appropriate to speak of a *permanent* technological-industrial revolution, which began with early industrialization, and in a number of phases (chemical, electrical, computer, cybernetic, nuclear technologies) has continued to gain momentum during subsequent times. If one focuses, however, on the *consequences* rather than the origin, and notes the radical changes in various areas of life that were initiated with the introduction of the industrial mode of production, then the characterization of this first phase as 'Industrial Revolution' is by no means an exaggeration.

While particular instances of artisan technology were relatively isolated and self-contained, industrial technology, by contrast, involved from the beginning a complementarity among inventions, which created an internal dynamic to technological development. Iron production (blast furnaces) and processing (rolling mills, lathes, machine tools), for example, made bridges possible and high-rise building which would never have been feasible if made from wood. The same applies to the steam engine — and later on the electric motor and internal combustion engine — which also would have been inconceivable without iron as the material of construction. Combustion engines, in turn, led to the utilization for new energy sources (coal, petroleum) which, when coupled with factory machines, made industrial production possible. Simultaneously the prerequisites were created for improved transportation (railroad, steamship), which in turn contributed to progressive industrialization (cf. Eugen Diesel 5.3, 83).

7. ENGINEERING SCIENCES AND NATURAL SCIENCES

As is the case today, technological activity of earlier times was largely determined by economic goals. Accordingly, technological knowledge was often treated in conjunction with economic questions. Johann Beckmann's *Anleitung zur Technologie* (A Primer for Technology), published in Göttingen in 1777, can be considered the first systematic study of technological procedures in Germany. It presents a survey and analysis of the artisan 'arts' utilized at that time in factories and manufactures. Beckmann was neither an engineer nor a scientist, but rather a professor of economics. He himself embodies the combinations of technology, administrative sciences, and economic trade science, which was not abandoned in Germany until the nineteenth century when technological schools, which treated problems of the engineering sciences in their own right, were established. This development had a certain parallel in France where in time the general engineering sciences freed themselves from the preoccupation with military technology which still existed at the founding of the École Polytechnique (Rapp *4.23*).

From a methodological perspective, the rise of the engineering sciences, to which modern technology owes its efficiency, is attributable to three factors:

(1) To the *scientific approach in the broader sense,* which consists in the use of (rudimentary) theoretical working hypotheses and empirical investigations to systematically order and critically test data about the respective procedures. It is obvious that by this process efficiency can be greatly increased. In contrast to artisan knowledge and ability, which is transmitted essentially by *oral* tradition, and thus dependent on the presence of a human mediator, in the engineering sciences technological knowledge, in the form of written codifications can be made generally accessible independent of particular persons.

(2) To the treatment of technological questions *independently* from other objects of study. With this clever device the entire attention is now centered exclusively on the structure of the procedures. In this way, the *specifically technological problems* can be identified and brought to optimal resolution. At the same time, it makes it possible to apply uniform and systemic criteria to the different areas of technology by means of a coherent overview. To be sure, this separation of the theoretical structure of engineering from questions of practical application is today frequently deplored. Specialization, however, is the unavoidable price for perfecting technological knowledge. In principle, this disadvantage may be overcome through the practice of interdisciplinary

cooperation. As to the decision about the use of existing technological means, however, grave problems arise as technological experts necessarily assume a key position.

(3) To the combination of technological knowledge and ability with the *mathematical-scientific* method, the most decisive factor. This connection, to be sure, suggests itself from both sides. Yet, from a historical perspective, it was not until the nineteenth century that the close connection between technology and the natural sciences, which is common today, became the generally accepted norm. It is immediately evident that we are dealing with reciprocal dependencies: in contrast to the qualitative and teleological approach of Aristotelian physics, which based its knowledge on passive observation of spontaneous natural processes, modern science is quantitative, mechanistic, and dependent upon experimental interference in the natural process. To establish the required (constant) experimental conditions and to be able to observe the results in an optimally precise form requires certain apparatuses including measuring instruments. From the outset modern experimental science has depended upon appropriate technological aids. Since modern-day research includes macrocosmic and microcosmic events, which themselves are only accessible with the aid of expensive equipment, the *dependency of science on technology* which has always existed has grown even more conspicuous.

In principle, the results of scientific research can always be used technologically. After all, technology attempts to create certain conditions in the material world which then lead to the actually intended results. Scientific experiments too are based on this same principle. In terms of subject matter, it makes no difference whether a physical or chemical result is produced in a research laboratory to verify a particular theory, or is created on a large scale to serve the goal of practical utility. The difference is only one of intention. The structure of the natural processes is the same in each case.

This does not, however, imply that all scientific knowledge possesses immediate practical utility. It is up to the engineer to *transform* these scientific principles into practical artifacts, while at the same time keeping in mind such practical design criteria as low cost, product safety, ease of maintenance, and dependability. The whole process is further constrained by the fact that a sufficient demand must be present to make production *economically worthwhile*. It is characteristic of technological practice today to systematically develop every important result of scientific research into an economical industrial product, in as short a time as possible.

Just as science has become more technological, so too has *technology become more scientific*. This is true of *content*, that is, the utilization of the results of scientific research for technology, as the preceding paragraph illustrates. In terms of *form*, it involves application of quantitative mathematical methods and formulation of comprehensive theories, which, through imposition of suitable initial conditions, lead deductively to the individual cases of interest. It is true, there are narrow limits to this procedure in actual practice. The primary task at hand is not, after all, to offer a theoretical *explanation*, but rather to *produce* an actual workable artifact. In contrast to scientific discovery, which consists in the demonstration of 'objective' cause/effect relationships in nature, technology aims at creating appropriate material structures based on subjectively conceived invention. In this process, intuition, know-how, and practical ability necessarily carry considerable weight.

While it is true that scientific investigation also relies on creative intuition, the role of mathematically formulated theory is much more significant than in the case of technology, where, for practical reasons, methods of approximation must frequently be resorted to. Nevertheless, the application of the mathematical-scientific method provides the only means of systematically increasing the efficiency of technological procedures. Thus, under the influence of the exact mathematical and deductive spirit of modern science combined with the knowledge of natural laws, the character of the technological invention has fundamentally changed. No longer, as was the case during the Renaissance, is invention attributed to a mysterious and spontaneous kind of higher intuition. This fact was perceptively expressed by Werner von Siemens, one of the founders of electrical technology, in the following terms. "Useful and practical inventions . . . do not really have to be sought for. They force themselves on one as the result of mature expertise and untiring work. Such inventions are solidly grounded on a questioning of nature by means of the experiment, which is informed through knowledge of her laws" (*4.19*, 812).

As has already been noted, science, an outgrowth of natural philosophy, has increasingly come to rely on technological devices in its pursuit of the experimental method. Conversely, technology, an outgrowth of the artisan tradition, has embraced science in its quest for increased efficiency. This relationship of mutual complementarity, which is grounded in the nature of the two fields, has only rather recently been fully developed to the point where it is now frequently impossible to distinguish between basic research in science and technology. How this came about is revealing. The expression 'engineering science' (*'science des ingénieurs'*) is first found around 1790 as the title of a handbook for construction engineers. Yet, as much as one hundred years later, mathematical treatment of technological tasks had hardly gained entrance into practice (cf. Rapp *4.23*, 20–21). The decisive stimulus for theorectical scientific research was often provided by practical problems in technology. A well-known example concerns the origins of thermodynamics. In an effort to increase the efficiency of heat engines, Sadi Carnot originated this discipline through his investigations into the ideal heat engine cycle. Eventually, thermodynamics evolved into an autonomous field within theoretical physics.

Like most other leading engineers of his time, Carnot was educated at the École Polytechnique, which had been established in Paris in 1795. In this school, which was to become a world-wide model for engineering education, for the first time a core curriculum was devised for all branches of technology which were subsumed under the name 'polytechnic'. In particular, this curriculum consisted of arithmetic, geometrical construction, and mechanics, with continuous alternation between lectures and practical laboratory exercises. With this approach, technology had, in principle, embraced the basics of scientific methodology: isolated and practice-based rules of experience that were characteristic of technology up to this time were replaced by a comprehensive theoretical approach which used mathematical methods and the results of scientific research verified by systematic experimentation.

It is important to contrast the polytechnic rationale with Beckmann's economics-based approach to the study of technology already referred to above. Beckmann took *areas of concrete technological action* as his point of reference, i.e., the crafts and trades, which were, for the time being, still studied in a context of economic considerations. From the beginning, however, the polytechnic view was based on *mathematics*, with technological problems studied specifically as to their abstract mathematical content. This approach, with its emphasis on methodology and formal mathematics and an exclusive concern with systematic exploration and efficient utilization of all technological possibilities, was to prevail in the times to follow. As such, it established the conditions for the rapid further advancement of technology which succeeded the first artisan-empirical phase of the Industrial Revolution.

8. INTELLECTUAL PREREQUISITES

Certain elements of technological development may only be

THE ROAD TO MODERN TECHNOLOGY 93

realized if corresponding social and economic conditions as well as the necessary state of technological knowledge and ability are also present. These prerequisites, which have been discussed above, specify the limits of technological potential, but do not automatically lead to particular technologies; they *license* a range of possibilities without *compelling* the realization of any particular instance. The collective process of technologically transforming the material world can take place only within the framework of these *external prerequisites*. Because man is, however, more than a mere creation of nature and because he participates in the 'realm of freedom', he *can* and *must* make decisions (individually and collectively) as to which way to use any given opportunity for action. His internal prerequisites for decision-making are comprised of specific motives, interests, or reasoning.

As that being capable of saying 'Yes' or 'No', man possesses the freedom to realize or to reject any factually possible technological action. Accordingly, to adequately understand the process which eventually led to modern technology one needs, in addition to a consideration of socio-economic and intrinsic technological preconditions, to pay attention to intellectual prerequisites. Even if one does not recognize the attitude out of which modern technology was born as being an autonomous phenomenon, but rather considers it only as the necessary result of 'real' or 'material' conditions, one must nevertheless concede that such prerequisites do exist *objectively* and, from a logical point of view, amount to *basic* prerequisites. The problem of a materialist interpretation of the development of technology thus consists in showing that these prerequisites are only derived and dependent.

Clearly, however, this should not be taken to mean that any selection process is totally arbitrary. Beyond "the will, wishes, and subjective intentions" the individual is influenced by the generally accepted views and "the categorial apparatus of thinking" of his times (Scheler *4.26*, 119). The development of technology and the intellectual milieu are always intertwined: the latter has been formed in the course of a historical process and in turn determines the future direction of technology.

Within the conditions of such a framework, each era defines its relationship to technology. Man's view of himself and of his world, results in concrete (technological) intensions which thus determine the way he will act vis-à-vis a range of technological possibilities. Each particular 'choice' he makes narrows the range of future options, without completely prescribing future developments. In terms of material conditions this is obvious, since every technological action depends on the available raw materials, instruments, and tools. But even in the case of mental attitudes, and in particular in the case of technological knowledge and ability, the process is never one of completely abrupt change, but rather one of gradual development. The present discussion is intended to show that the philosophical views, technological practice, and research in the natural sciences have all contributed, in a complex and intricate process, to the present situation.

The specific character of modern science-based technology can be better understood if one focuses on the intellectual prerequisites for its origin. To facilitate clear description, the following will emphasize *systematic* coherence more than *historical* genesis: In retrospection on the entire process, the intellectual prerequisites relevant to the rise of modern technology will be investigated. Eight such preconditions will now be discussed.

a. Valuation of Work

All technological activity is based on intervention into the natural process and therefore on *physical work*. Admittedly, modern technology contains an extensive intellectual component. However, any systematic, large-scale technology such as ours could develop only where manual work was recognized, in principle, as valuable. This was not the case in the classical period. The Greeks considered physical work as inferior, since it was necessitated by physical needs and neither served politics nor was connected with theoretical knowledge, which is characteristic of man as *animal rationale*. Physical work was relegated to slaves and low-ranking, uneducated artisans. This attitude is probably the reason why, despite high intellectual achievements in classical Greece, no large-scale technological projects were undertaken. The high valuation of theoretical knowledge and the corresponding low appreciation of practical technological work survives in various forms even today. Technical colleges in Germany, for example, were not permitted until 1899 to grant the Ph.D. as a recognition of their scientific status. Even then such recognition was gained in the face of vehement university resistance. Under the influence of Christianity, the value of physical work gradually changed. It is true, in the Old Testament labor is considered as a curse, and the New Testament primarily demands inner contemplation and not external activity. Yet by the Middle Ages, the statutes of the monasteries frequently prescribed regular manual work. Mumford even thinks that it was precisely the ideal of ascetic life of the monks along with their concentration on interpersonal tasks that prepared the object-related attitude which characterizes modern science and technology (*4.21*, 34). During the Reformation, work and the execution of vocational duties underwent a clear increase in valuation. Thus, especially in Calvin's doctrine of predestination, incessant work and vocational success serve as a sign of genuine belief and divine selection.

With the Protestant abolition of monasticism, vocational life in the world became the realm of Christian asceticism. According to Max Weber, it was this religion-based ascetic work ethic, demanded of entrepreneur and worker alike, which served as one of the prerequisites for the rise of modern capitalism (*4.29*, 174–179). In Protestant regions, one can indeed generally observe a stronger striving for professional advancement along with greater economic activity, than in Catholic areas. Even after the loss of the religions foundation, this ethical attitude towards work and profession survives, to a large extent. It is typical, however, of the present affluent society of West Germany, in which individual demands continue to grow, that this achievement principle is weakening. On the other hand, in East Germany the ascetic tradition has been secularized or reinterpreted and consciously cultivated. Indeed, functionality and productivity of a rigid technological working world requires such a concept of vocation. Such an attitude toward work is by no means a matter of course, as evidenced, among other things, by the difficulties involved in introducing efficient, technology-oriented work styles into developing countries in which spontaneous life styles and work behavior are common.

b. Efficient Management

A further prerequisite for modern technology is a system of *efficient management*. Here as well, Weber sees a connection between disciplined and methodical styles of personal life and of economic processes, on the one hand, and the spirit of Protestant work ethics, on the other: Puritan-ascetic concentration on vocational achievement requires sober and continuous work and prohibits both the wasting of time ('time is money!') and unnecessary rest. Pleasure-seeking and waste are reprehensible; only practical things are worth striving after (*4.29*, 155–172).

In keeping with his doctrine that earthly success is a sign of predestination, Calvin explicitly declared in 1545 that the charging of interest was legitimate. While it is true that before this time there were money-lenders (usurers), their dealings were generally regarded as immoral. It was only through the religions authority of the reformer Calvin that striving after profit and acquisition as a legitimate form of one's vocational calling could be reconciled with the command of brotherliness (Hoselitz *4.12*a, 63–65). Thus *traditionalism*, which is characteristic of agrarian structures and aims at the preservation of established conditions, is supplanted by *economic rationalism*, which is typical of city life with its consciously and purposefully calculated norms (*4.12*b, 658). Out of the Protestant "mentality which *vocationally* strives for legitimate profit ... in a systematic and rational way," the corresponding organizational forms, such as exact calculations based on bookkeeping and efficient management, were generally introduced (Weber *4.29*, 64–65).

c. *The Impulse for Technological Creativity*

In a comparison between typical values and lifestyles during the Middle Ages and in modern times, a fundamental change is noticeable, even though it is impossible to say where one age ends and the next one begins: the Middle Ages, like preceding eras, are, in principle, characterized by a static and conservative image of the individual and society. A stable, well-defined social structure, together with a demand-oriented economy (whether feudal agriculture or city-based trade), in combination with a fixed framework of religion created a strong conservative tendency, resistant to change and preserving of traditional organic institutions. In contrast, modern times are characterized in all areas by a constantly intensifying creative impulse, which finally reaches the point where 'change' itself becomes a value. Additional sources of this modern attitude include the secular orientation and high valuation of creativity of the Renaissance, the abolition of traditional authority, and the reliance on autonomous rationality. Without such an impulse and the resulting willingness to fundamentally change the entire situation, the speed of development of industrial technology would not be conceivable. As Sombart notes, this modern spirit is characterized by a "striving for infinity" which continuously sets new goals and does not recognize natural limits (*4.27*, 327–328).

This active attitude stands in necessary opposition to a contemplative lifestyle, with its emphasis on inner harmony or reflective pleasures. Because unceasing activity makes impossible the joy of achievement, people find it difficult these days to actually appreciate the 'better life' technology was, after all, supposed to have provided: a case in point is the expression 'coping with leisure time' (*Freizeitproblem*).

As a matter of fact, when viewed from a historical perspective, it is clear that the exaggerated 'will to technology', characteristic of modern man has by no means always been present and cannot be regarded as a natural anthropological feature. Consider the Chinese tale from the third century BC of an old gardener who through long years of work had carved steps into a rock wall and daily used them to carry water from a well. When it was suggested that he might lighten his task with a draw-well he rejected the idea. "I know this invention," was his reply, "but I would be ashamed to use it" (Rüstow *4.25*, 63). A similar spirit is reflected in the response of an Oriental cadi to a European researcher requesting information about the city of Mossul: "What you ask of me is at the same time useless and harmful . . . If you think that these scholarly things have made you a better person than I, then be twice as welcome to me for it; I, however, thank Allah that I do not search for things that I do not have to know; you are knowledgeable about things that do not matter to me, and what you have seen, I despise it. Will your more comprehensive knowledge create a second stomach for you, and your eyes which glance everywhere and ransack everything, will they find paradise for you?" (quoted from Peter Mennicken *4.20*, 152). The unconditional clinging to traditions, and the respect for a sacred natural order which these tales exemplify is completely alien to us nowadays. Precisely in this contrast, however, is underscored the fact

THE ROAD TO MODERN TECHNOLOGY 99

that the modern attitude of being afraid *not* to apply a technological innovation is itself idiosyncratic to a particular line of development of the history of ideas.

In the same vein, Hall appropriately points out that at the stage of artisan technology, while many persons were daily using looms, mills, plows, horse-harnesses, for centuries not even minimal technological innovations were introduced, although, from our perspective, this would have demanded no ingenuity. Rapid technological development began only when machines and more *complex* technological procedures were employed involving only relatively few persons directly, and when inventions came to require increasingly more detailed knowledge (*4.11*, 128). This shows that the desire for technological progress presupposes a specific predisposition sustainable only in a particular mental climate. It is thus Western history of culture and ideas that has provided the intellectual prerequisites for the large-scale transformation of, first, Europe and, then, the world.

d. Rational Thought and the Enlightenment

One further decisive intellectual prerequisite for systematically rationalized modern technology is *rational thought*. It evolved in the course of European intellectual history and has become the vehicle for technological creativity. Especially with the *Enlightenment* ideal of universally valid, logically conclusive, and personally verified knowledge, this attitude confronted naive every-day ideas as well as unchallenged traditional religious or scientific authorities. Such object-related rational thought aims at intellectual understanding for its own sake, while excluding subjectivism, and abstracting from the concrete context of everyday life and history. It culminates in the idea of 'value-free' science.

In the *theoretical field* this non-partisan and ascetic attitude regards objectivity and universality of scientific understanding as its supreme goal. Applied to *practical action,* the same style of thinking offers the possibility of 'optimal' solutions to technological problems, which are, by their very nature, 'value-free', impersonal, and interchangeable, and thus themselves the very criteria of objective thought. The rational and prosaic — and for that reason so successful and efficient — concentration on technological procedure leads, however, to the necessary consequence that the intellectual energy is entirely directed toward impersonal and thus 'inhuman' questions. Certain tendencies of the times toward irrational escapism or unreflective and purely emotional involvements can be understood as a reaction to the cold anonymity of the modern technological and scientific world.

The origins of Rationalism lie in Greek antiquity. By way of a summary survey one could even consider the whole of European history of ideas as a continuing process of enlightenment, reaching into the present. This development, nevertheless, was by no means straight and without breaks. A decisive turn was taken with the transition from the Middle Ages to modern times, a period generally interpreted as a process of secularization, i.e., as emancipation from clerical tutelage and orientation toward this world. This interpretation must, however, be qualified in certain respects. It was, after all, the medieval Church which realized, in its constitution, its legal structure, and particularly in scholastic philosophy, essential elements of rationality (Brunner *4.4,* 91). Just as in Marxist theory, which follows Hegel's historical teleology, the capitalist mode of production appears as a necessary transition state on the road to communism, so too the rationalistic tendencies of the Church-dominated Middle Ages could be considered as the prerequisite for the secularized thought of the Enlightenment.

In contrast, such a view, according to which the modern era takes shape only against the backdrop of the Middle Ages, Hans Blumenberg maintains the "legitimacy of the modern period." For him, modern science is an instrument for the self-assertion of reason which itself needs no additional justification. In industrialized technology he sees "man's self-assertion countering nature's inhumanity" (*4.3*, 265). If, on the other hand, one follows Arendt, then the modern times, which are dominated by technology, are characterized precisely by the fact that reason, in the sense of an originally given and self-revealing contemplative understanding, is *lost*, to be replaced by a detached (technological) activity preoccupied with abstract mathematical theory and physical replication (*4.1*, 248–304).

These two opposing positions express the *ambivalent* character of modern technology: It has created the prerequisites for a life largely liberated from the insufficiencies and limitations of man's biological nature and can therefore be understood as appropriation of nature by self-defined, autonomous man. At the same time, however, modern technology has created new constraints and risks for cultural and intellectual existence, and even for the physical survival of mankind. The evaluation of this process will vary depending on which definition of 'reason' is used. If reason is interpreted in the sense of rational mastery of problems, then the autonomy and resulting transformation of nature that it makes possible is to be evaluated as a positive achievement. Beyond this definition, however, reason can also function as the basis of authority for contemplative reflection and a meaningful personal existence. On this score, modern technology must be assessed in less positive terms.

e. Objectification of Nature

Today's technology owes its efficiency to the systematically calculated use of natural processes. According to present understanding, the physical world is an arbitrarily manipulable object: disregarding the state of material and intellectual resources, the only definite limitation of technological activity is given by the laws governing the respective physical processes. As the history of culture and ideas shows, however, this attitude is by no means a matter of course. It evolved only in the course of a 'de-sacralization' of nature along with an abandonment of magical beliefs. In the end, nature was no longer understood as animated and worthy of reverence, but instead came to be regarded as an accidentally present and arbitrarily useable conglomeration of lifeless matter. Thus, *objectification* of nature constitutes one more prerequisite for the systematic transformaion of nature through modern technology.

Before reaching this final position, however, one should point out an intermediate position according to which the cosmos appears as a sacrosanct, organic whole, in which man, too, is integrated as part of nature. Such a position abandons an appeal to magical thought and the concepts of the alchemists, yet offers religious reservations against a purely 'materialistic', objectified view of nature. This can be seen particularly clearly in Francis Bacon who — far ahead of his own times — propagated the idea of a scientific technology. He could justify his position only by pointing out explicitly the God-willed and morally unobjectionable character of systematic natural research which, in his opinion, was by no means to be equated with the hubris of the Fall (*4.2 IV*, 20). Similarly, the authors of the "machine books" of the 16th through 18th centuries, without exception, referred to the concepts and ideals of the Biblical-Christian tradition in support of the concept of *ars mechanica*. The attempt to show the compatibility of technology with Christianity, however, has increasingly led to a reinterpretation of the Biblical tradition in favor of technology: "Work creates the Paradise; the miracle should seem as much a natural event as possible." On the basis of this reinterpretation and assimilation, technology was able, in the times to follow, to be generally accepted without serious challenge (Ansgar Stöcklein *4.28*, 111–112).

Though largely secularized, an organic and holistic view of nature is still characteristic of the Romantic Age and – with individual modifications – of the positions of Goethe and Alexander von Humboldt. With the success of the Industrial Revolution and 19th century scientific research this view finally vanished from the general consciousness. It is only in the areas of esthetic experience, leisure-time encounters with nature, and calls for preservation of the natural environment that it survives in rudimentary form. The gradual abandonment of an organic-holistic view obviously signifies not only *progress* in scientific knowledge and *gain* in technological capacities but also *loss* in the direct and self-evident experience of nature (Arendt *4.1*, 250–266). This shows once more that there is a price to be paid for the accomplishments of modern technology; they can be realized only if the world is considered as an arbitrarily disposable inventory for technological uses that are permitted by the laws of nature. All esthetic, ethical, and existential aspects of nature are thus excluded in principle from scientific technology. Any reintroduction of such 'foreign categories' therefore requires additional justification.

f. The Mechanistic View of Nature

One further prerequisite for modern technology is the *mechanistic view of nature* (of Eduard J. Dijksterhuis *4.7*). This concept, already hinted at by the ancient atomists, does not become generally accepted for the description of natural processes until the transition from medieval to modern times. For persons unfamiliar with mechanical apparatus, the concept of nature as a gigantic machine is by no means natural or even plausible. Children, for example, although they may grow up in a mechanized environment, need years of instruction to became familiar with the mechanistic mode of thought. On the other hand, the elementary experiencing of nature, along with biological growth processes and goal-oriented human actions is much closer to the teleological perpective of Aristotle. This explains the orientation of natural philosophy and natural science up to modern times. Within the framework of this finalistic view of nature an obvious distinction is made between natural, spontaneously occurring processes and artificially initiated ones. The mechanistic viewpoint alters the teleological approach in three essential respects:

(1) The root metaphor is no longer naturally occurring processes, but rather *mechanical* processes, *artificially* generated with the aid of appropriate apparatuses and instruments.

(2) The conceptual inventory for the description and analysis of all natural phenomena is no longer obtained from the 'higher' and complex *organic* processes, but rather from the 'lower' and simpler *inorganic* processes.

(3) The *synthetic* and teleological perspective, which focuses on the final result of processes, is now replaced by the *analytic* investigation of the *functional* relationships between spatially and temporally contiguous states of affairs.

The mechanistic view of nature, on which modern science and technology are based, results both from concrete experiences and theoretical considerations. Precursors for both constituents can be traced to the late Middle Ages. The theoretical conception is based in particular on the physical impetus theory and on the beginnings of a quantitative mathematical description of natural processes within the framework of scholastic philosophy (Dijksterhuis *4.7*, 187–200). The empirical foundation for mechanical thinking stems mainly from the elaborate clock constructions of the late Middle Ages. In the 14th century, Nicholas Oresmus was the first to interpret the universe as a gigantic mechanical clock which, as he saw it, had been created and set in motion by God in such a way that all its parts turned in optimal harmony (White *4.30*, 125). The philosophical justification and authoritative formulation of the mechanistic worldview was provided by Descartes (cf. Arno Baruzzi *1.5*). This combination of mathematical formalism and the mechanistic view of nature provided the basis for classical mechanics as originated by Galileo and refined by Newton, and has further served as the foundation of all subsequent scientific theories. Since both science and technology are based on the mechanistic view of nature, they support each other. This process of mutual stimulation has only in present times developed its full momentum.

In previous eras, technological activity was, to a large extent, embedded in the organic-biological sphere (hunting, fishing, agriculture, raising cattle, riding-animals and beasts of burden). Modern technology, on the other hand, owes its efficiency to the clever device of having severed these limiting connections with the organic sphere. Now the desired technological effects are achieved by means of appropriate artifacts, through artificially induced inorganic-mechanical processes. When compared with biological evolution, which proceeds from simple inorganic forms to more complex organic structures, modern technology is in one sense regressive: it takes recourse to the more primitive and more easily handled processes of the inorganic state, an earlier stage in evolutionary history. The systematic unfolding of this procedure which, strictly speaking, is 'unnatural' for man as a biological being, has made it possible to far surpass the practical achievements of organic technology.

g. *The Mathematical Model*

A further condition for the efficiency of modern technology is the use of *mathematical methods*. The determination of the dimensions of technological systems and of the magnitudes of relevant physical and chemical processes require appropriate calculations, which lead, at least in approximation, to the optimal technological solution. The exact quantitative description of technological processes is historically and systematically in close relationship to the mechanical view of nature. It is not, however, identical with it. Consider, for example, the fact that simple mechanical apparatuses can be built without mathematical calculations, solely on the basis of qualitative concepts and traditional artisan knowledge. Conversely, the history of mathematics shows that abstract calculi, formalistic and logical in character, can, to a large extent, be developed without reference to concrete application. On the whole, however, the close relationship and cross-fertilization of the mechanistic view of nature and mathematical methods is, for modern times, an obvious fact.

The mathematical approach has become important for technological development in three respects:

(1) The theoretical ideal of *quantitative exactness*, which has its empirical counterpart in the exact measurement of observational variables, makes it possible to compare competing procedures. This creates the prerequisite for a technologically optimal use of available resources.

(2) The concept of *functional dependency*, the simplest case of which involves one independent and one dependent variable, supplies a conceptual repertory for the isolated investigation of specific relationships, and makes it possible to fully exhaust the technological potential implicit in the relationships.

(3) Operating with *abstract variables* provides a starting point for the conceptual and the concrete isolation of individual elements, processes, and phenomena. When these are *combined*, an entire realm of new technological possibilities is generated.

h. Experimental Investigations

The mechanistic view of nature and the mathematical method supply theoretical potential. They do not, however, guarantee practical applicability. They always require verification by concrete data in order to ensure that they do indeed describe actual physical phenomena. Methodologically, the rise of modern technology and of the exact sciences is, thus, based on the *combination of mathematical-mechanistic theory and systematically performed empirical investigations*. What is actually new about this is the active intervention in the natural process in contrast to the premodern custom of passive observation. Only this goal-oriented 'interrogation' of nature by experiment, and with the aid of appropriate apparatus, made it possible for man to systematically use the physical world.

This attitude of active outreach is itself a specific instance of the objectification of nature mentioned above (point *e.*): In all experiments, nature is considered as an object that is completely at our disposal. The basic theoretical attitude underlying this approach was formulated by Bacon, who compares empirical investigations with a court trial which has been granted to mankind through divine mercy and providence, in order that we may claim our rightful authority over nature (*4.2 IV*, 263). Kant repeats the metaphor, though in a secularized version, in observing that reason approaches nature by means of the experiment. "It must not, however, do so in the character of a pupil who listens to everything that the teacher chooses to say, but of an appointed judge who compels the witnesses to answer questions which he has himself formulated" (*4.14*a, B XIII).

Technology and experimental investigation are closely related. Each produces objects or processes which are either useful or interesting. One can view all technological constructions (successful or unsuccessful) as experiments leading to particular insights. Only by means of precisely controlled experiments, however, is it possible to isolate and examine in detail the relationships that are of interest. The 19th centrury creation of the engineering laboratory not only contributed significantly to technological progress, but provided a source of new possibilities for experimental investigations themselves.

The systematic experimental approach, guided by theory and executed with the aid of technical instruments, has provided a *welcome* empirical foundation for scientific knowledge and technological action potential. At the same time, however, the approach has made it possible to manipulate the physical world in an almost unlimited range of ways, some of which open up extremely *dangerous* horizons. The trend has been to fully realize each technological research potential. Should such a trend continue, mankind will edge ever closer to self-destruction through constantly refined super-weapons or by means of genetic manipulation. Because the physical world sets no limits to theoretical investigation and practical exploitation − except those immanent in its own structural laws, ever more accurately comprehended by modern research − it is up to mankind to deal with nature in self-limiting ways if a catastrophe is to be avoided.

9. COMPLEX INTERCONNECTIONS

Discussion of the historical interrelationships and systemic interdependencies between the socio-economic, technical and intellectual prerequisites of technology has demonstrated the inadequacy of applying a linear or single-cause explanation to what is actually a complex and convoluted process. In particular, the complete reduction of the historical process to either intellectual factors (theoretical interests, value orientations, philosophical presuppositions) or to material factors (production processes and materials or socio-economic conditions) constitutes an inappropriate simplification of actual events. Each extreme overlooks the fact that, although philosophical views and materials are each components of relative *coherent historical processes*, nevertheless both components always interact to produce the initial conditions of any given time. While, as the existing literature shows, it is possible to write a self-contained history of technology from the

THE ROAD TO MODERN TECHNOLOGY 109

perspectives of socio-economics, engineering technology, or history of ideas — on the basis of a disciplinary division of labor, certainly a legitimate approach — to ignore the interconnectedness of all such perspectives can lead to a kind of disciplinary myopia. Because historical data can only evidence the *simultaneity* of particular phenomena, the actual problem is one of explanation: is the connection between the phenomena one of *genuine interaction* or of *cause and effect*? An action theoretical analysis will always yield several relevant factors: the technology is performed by an individual or collective *acting subject*; it occurs in a concrete historical *situation* containing socio-economic, technical, and philosophical factors; finally, it is intentional or *goal-directed*. On the one hand, the intellectual schema must reflect material conditions: socio-economic and intrinsic technological factors. On the other hand, however, material conditions do not by themselves dictate certain technologies; it is only on the basis of a specific *evaluation* of the situation that goals emerge. Past events always define a range for action within which decisions concerning the given technological potential must operate.

An appeal to the idea that individual or social needs can clarify this problem does not have self-evident content. The desire to relieve a state of tension, or to supply some need, is not simply given by nature in its concrete form. With the exception of elementary biological reflexes, such a desire can only arise within a system of value preferences which assigns positive or negative values to *certain states* that are attainable. This line of reasoning suggests that it is the cultural-historical relativity of value systems, and not socio-economic conditions which are themselves dependent on particular value orientations for their existence, that explains the variations of developmental patterns of technology among various high cultures.

From this it does not follow, however, that modern technology is the result of a completely arbitrary and quasi free-floating change of values. As the sketch of historical development has shown, the process which ultimately led to the present situation has exhibited something of an inherent logic, itself characterized since the Renaissance and particularly since the Industrial Revolution, by mutual dependency between technological conditions and value concepts. It was technology itself, for example, that generated or actualized a need for luxury items to the point that they are now accepted as a matter of course in the industrialized nations. Kenneth Boulding suggests we are faced here with the chicken and the egg problem: that is, it is generally impossible to determine whether new technologies produce new values or vice-versa (*5.2*, 347). This opinion, formulated with an eye on the present situation, which is dominated by technology, gains its plausibility from the fact that any value concepts reflect the existing environment, which is always shaped by technology as well as that time's (technological) action potential. Nevertheless, future developments are not inexorably determined since we can reassess values, alter our attitudes, and replace a blinkered acceptance of technological developments with a regard for future generations.

10. NATURAL INSTINCT AND VOLITIONAL CREATIVITY

Although the establishment of complex historical interdependencies provides valuable insight, it does not satisfy the yearning for a unified and lucid interpretation of the development of modern technology. There is, consequently, no lack of attempts at identifying, by means of abstract frameworks, one fundamental principle lying 'behind' concrete action processes. These approaches are obviously based on the expectation that it is possible to find a monocausal explanation couched in general and basic terms. It should be noted in this connection that there is frequently in the literature no clear distinction made between a *methodological* concentration on a single aspect, which may then be investigated

THE ROAD TO MODERN TECHNOLOGY 111

in detail, and a more radical approach that assumes that treatment of this one aspect can provide an exhaustive explanation of the technological process *in toto*.

In this regard, two basic positions as to the origin of technology may be distinguished. Roughly characterized, they are the practical-naturalistic and the theoretical-cultural approaches. The practical-naturalistic interpretation of Gehlen is based on the premise that human beings are organically "deficient", lacking both requisite fixed instincts and other natural endowments necessary for survival in a hostile environment. They must depend, accordingly, on the production of tools and prudent, planned, and specialized (technological) action to modify nature in accordance with human needs (*5.9*, 46—48). Technological intervention in the natural process and the production of artifacts are thus not to be explained as mere examples of rational thought or considerations of utility. To the contrary, they represent indispensable stratagems by which mankind "relieves" the pressures generated by the demands of physical life. Mankind is thus intrinsically dependent upon technology (94).

In Gehlen's opinion, clearly marked processes that are repeated over and over and take place automatically are crucially important, since they respond to the need for environmental stability and offer a substitute for the lack of instincts. The fascination generated by well-functioning mechanical devices (such as clocks and motors) is based on some intrinsic attraction to technology which is itself quite distinct from considerations of such devices solely in terms of purpose or utility. For centuries, for example, effort has been expanded in constructing a *perpetuum mobile* (which would function continously without an external energy source) without pursuing practical purposes with it. The instinctual process of mechanization rooted in human nature is based on closed 'circles of action', i.e., action processes which are controlled by the feedback of success or failure achieved in each case and which ultimately change into routine automatic movements, thereby relieving man of the task of having to master each action situation anew (*5.8*, 13–19). In summary, Gehlen concludes that "technique, from its beginnings, operates from motives that possess the force of unconscious vital drives. The constitutional human features of the circle of action and of facilitation are the ultimate determinants of all technical development" (19).

The real achievement of Gehlen's anthropological approach lies in his drawing attention to non-rational determinants of technological action as opposed to purely philosophical or intellectual explanations. It is, however, debatable whether Gehlen's approach is as comprehensive as he claims. Man's relief from physical work through simple mechanized processes may indeed reflect an innermost instinctual need. Yet, it is a long way from organ substitution through the use of tools and from performance of elementary engrained and reflex-like actions to the highly complex systems of modern technology. Furthermore, the fascination with functional and intelligible mechanical processes finds its counterpart in the frustration stemming from technical processes that defy our efforts to understand them. And the monotonous and isochronous character of mechanical processes stands in stark contrast with the flexible rhythms of nature governing man's biological existence.

Thus the present impulse toward a radical technological transformation of the world can be interpreted as the result of an initial natural drive, but not in its concrete shape, as it is observed today. If, following Gehlen, one were to reduce to a constitutional, instinctual need the historical fact of man's efficient appropriation of nature, one would need to include, in addition to the positive (or at least neutral) relief instinct, a negative and destructive impulse for technological action as well, given the ruthless exploitation of nature and varied forms of technology-related alienation that exist. In this case one might arrive at some conception such as that of Ludwig Klages, who sees in 'Intellect as Adversary of the Soul' (*4.15*) the constitutive principle of technology. According to the direction of Gehlen's anthropology, the strength of his investigations lies in the explanation of why mankind took the road of technology *in the first place*. His theory is more likely to explain flint than the atomic bomb; it shows the origins of technological action, but fails to elucidate the structure of its historical development or its increasing momentum.

To a certain extent, Gehlen's central concept of relief can correctly describe the process of intensified technological development as 'delegation' of labor to technological artifacts. The motives for this specific kind of dealing with the material world, however, can hardly be based on universally given endowments of human nature, since in that case, development of technology would have had to be identical everywhere in the world. To a great extent, this was indeed the case during early prehistoric times, but with the development of artisan technology considerable differentiation began to occur. And when it comes to modern technology, a phenomenon *sui generis* in the total picture, any explanation requires an appeal to specific causes.

Gehlen's hypothesis of "relief" (*Entlastung*) does not supply a satisfactory explanation in this case, for it assumes the existence of a biologically caused *pressure* (*Belastung*) which requires technology for its relief. Such a nature-induced crisis, however, is *not* present in the specific case of modern technology. To be sure, the construction of supersonic planes, nuclear reactors, and color television sets constitutes, among other things, relief from certain physical activities, however these technological aids do not serve the purpose of averting an actual danger. Once biological survival is insured, it requires a *cultural* impetus, which aims beyond basic necessities of life, if one is to perceive a given state as requiring change, and to develop further-reaching 'needs'. Thus *today*, it is true, since automobiles, trains, and airplanes are available, any extended travel on horseback or by stagecoach would be considered an imposition, but Goethe, on his journey to Italy, could not have missed these modes of transportation any more than we can feel deprived of technological innovations still to come. A comprehensive analysis which is to do justice to the concrete development of technology must therefore consider, in addition to the instinctual determinants, also the intellectual prerequisites of technological action.

Biological drives and cultural factors need not be construed as mutually exclusive determinants of the process of technological development. A conception that posits a complementary relationship has a better promise of appropriately describing the actual situation. Thus, one could start from the assumption that the excess of instinctual drive, always present according to Gehlen, has, in the course of historical development, increasingly concentrated on technological actions. Since the beginning of the modern age, and in particular since the beginning of the Industrial Revolution, the efficiency of the artifacts produced and of the procedures applied have, due to rational management and use of scientific methods, increased in a heretofore undreamed of manner. In combination with the structure of accumulation inherent in technological action, this process has led to a permanent 'self-reinforcement' of technology. This approach will, as a result, explain why the natural impulse for technology developed in a particular fashion due to cultural conditions that evolved during the history of Europe, and how the continual increase in technological potential, together with the dynamics of expanding needs led, in the following phase, to the explosive development of technology.

In the same vein, Arendt points out that "if we had to rely only on men's so-called practical instincts, there would never have been any technology to speak of" (*4.1*, 289). For her the origins of technology lie not in the perfection of tool usage, but rather in the striving for *theoretical* knowledge in the framework of experimental scientific research as founded by Galileo. This opinion may not agree with every detail in the history of science and technology, but this much is true, the present state of technology could not be imagined without the contributions of purely theoretical investigations, which while aimed at scientific understanding, contain, in principle, the potential for practical application. Yet, technological potential thus created, and supplemented these days by applied technological research, can only be actualized in proportion to the public's readiness for technology. This fact points up the way spontaneous and volitional elements come to bear on technological action.

It could be objected that there actually does not exist any choice at all, since technological development is totally determined by particular *economic* conditions. If such a position is adopted — as is the case with, e.g., Marx (*4.17*a) and Fr. von Gottl-Ottlilienfeld (*2.1*) — then the satisfaction of material needs in the economic process assumes the function of ultimate cause which Gehlen assigned to the biological constitution of man. Like the individualistic and anthropological interpretation, however, the socio-economic explanation, taken by itself, fails to be convincing. Each approach reduces the development of technology to an original natural process, while regarding cultural value systems and achievements of scientific theory as mere dependent variables. In contrast to such naturalistic interpretations of technology, Manfred Schröter emphasizes the "elements of creativity, self-reliance, self-fulfilment, and personal satisfaction in technological activity", as expressed, for example, in the pleasure which accompanies the creation of technological devices and the urge to play with them. Furthermore, he emphasizes that from the beginning technology has been in the service of culture, for example, through construction of sacred buildings and funeral sites (*1.26*, 25).

Cassirer, too, advocates a more comprehensive understanding of technology. He points out that in a methodological analysis technology is described only in the concrete terms of the sciences. Such a confining perspective, however, loses sight of the fact that technology, like language, art, religion, historical tradition or science is always a living part of culture and thus participates in the "totality and universality" of intellectual life (*1.9*, 21). The essence of technology, which, Cassirer holds, is independent of all external purposes, lies in a universal impulse to form manifesting itself in concrete acts. Technology, like language, is not some inert object. It can only be grasped as one manifestation of the force of dynamic creativity, one whose function and orientation is technological (24).

This thought addresses a central problem of modern technology; precisely because technological activity is always inseparably connected with the cultural life of society, there is a growing tendency, in view of the technologically shaped external conditions of life, to subordinate patterns of thought and general views of life to technological conditions and the so-called 'inner logic' of things. Depending on one's standard of evaluation, a world dominated by technology appears as a goal full of promise and an earthly paradise or else as a menacing prospect characterized by a loss of culture and humanity.

The range of conceptions mentioned indicates the complexity of the phenomenon 'technology': the need to cope with the environment and the existence of corresponding natural-biological drives are just as real as is the fact that higher forms of technological activity belong to the realm of intellectual and cultural creativity. *On the whole*, then, despite many transformations and interruptions, there has been a continuous line of development leading from the most rudimentary techniques of 'primitives' to the complex systems of modern technology. The fact that technology is ultimately rooted in the human constitution explains

THE ROAD TO MODERN TECHNOLOGY 117

why history portrays man at every stage as *homo faber*. And, since technology is part of the sphere of culture, it has an *open horizon* and can, like all other forms of cultural achievement, manifest itself in a multitude of changing forms.

CHAPTER V

THE TECHNOLOGICAL WORLD

1. NATURE AND ARTIFACTS

In contrast to natural processes, which are untouched by human intervention, all technological systems and processes are 'artificially' generated objects, or artifacts. This is why technological creations, in contrast with the first, autogenous nature, are frequently referred to as 'second' nature. Nevertheless, it would be a mistake to speak here in the strict and comprehensive sense of an anti-nature, or counter-nature. Closer inspection reveals at least three distinctions that must be made:

(1) All technological projects can be realized only to the extent that they are compatible with physical laws. Regardless of whether the applicable natural laws are completely known and explicitly formulated or are taken into account only through intuitive rules of experience, technological actions are always dependent upon the immanent physical and chemical cause/effect relationships. Here Bacon's *"natura non nisi parendo vincitur"* applies: technological action is based on obedience to nature (*4.2 I*, 157). Unlike social history where there have been cases in which outmoded power structures totally deprived their subjects of personal freedom, turning them into complacent objects of manipulation, the natural process can never reach the point of infringing on nature's autonomy. The clever principle of technological activity consists precisely in not forcing unrealistic goals upon nature – in contrast to magical thinking – but instead adapting as accurately as possible to her own immutable laws. Within the realm thus obtained – and within the limits of the existing resources – man has

complete freedom to direct natural processes according to human goals. All technology created so far, as well as all that will be created in the future, is thus in agreement with the laws of nature and is to this extent not at all unnatural.

(2) In terms of shaping the material world, modern technology is by no means a unique phenomenon. Even inorganic structures and processes possess some sensibly perceptible form rather than being formless and amorphous. Indeed, organic nature, apart from man's action, is structured as a self-contained and variegated order by the manifold forms of flora and fauna. Man, who is by his biological nature dependent on technological action, has always interfered with the natural balance of these organic processes, modifying them according to his objectives. If virgin nature, completely untouched by the human hand, is taken as the point of reference, then a flint is as unnatural as a nuclear reactor, since both are based on human intervention in the existing environment. When considering man in the context of evolution as a biological being, his technological products are even comparable with the constructions of animals which are generally regarded as 'natural' (beaver dams, termite hills, birds' nests).

(3) Finally, one must bear in mind that the definition of a natural state, to which we refer in judging technology as artificial, is itself subject to change through time. This is only reasonable since man has always existed as *homo faber*: each technological innovation is thus of necessity based on a preceding state of technology. From this perspective, the first step toward technology action can also be considered as the first step toward humanization (*Menschwerdung*). As Serge Moscovici emphasizes, when farming is initially considered natural, this is in contrast to the 'artificial' work of the artisan. Yet, with ensuing developments, artisan work comes to be regarded as natural in contrast to the artificial constructions of the engineer. Rather than evaluating technological development from successively changing viewpoints, Moscovici

regards it as a continuous process for which there exist no absolute criteria for distinguishing between natural and unnatural elements of the man-environment relationship. Thus, each technological *milieu* created by man is, in the last analysis, just as natural as every other one (*1.24*, 42).

Up to a point, there is a definite appeal to this argument. It is flawed, however, in that it overlooks the fact that man as a biological and cultural being cannot limitlessly adapt to self-created changes in his environment. An excessive rate of change can lead to harmful consequences. In coping with technology, mankind has shown an amazing ability to adapt to changing situations. It is an open question, however, whether individual self-definition and societal structures can really master the technological world in which change is ever accelerating. Since technology originated in the industrial nations, their cultural traditions accept technology as indigenous rather than rejecting it as a foreign body. For the developing countries, however, with their different historical development, the problem of adaptation to rapid technological change presents itself in unmitigated sharpness.

As this point serious differences of opinion arise. On one side, authors such as Arendt (*4.1*), Ellul (*5.4*), and Hermann J. Meyer (*1.22*) reject technology in favor of a concept of nature and life determined by the traditions of artisan technology and by a lifestyle that is close to nature. On the other side are those such as Stanisław Lem (*5.23*), Moscovici (*1.24*) and Ribeiro (*4.24*) who visualize, in an optimistic embrace of technology, unlimited possibilities of change and of human adaptability. Gehlen, too, tends toward the latter position, and points out that mankind was able to master the critical transition to industrialization as it had already mastered the transition to agriculture (*5.8*).

Such theories, however, are based on more than bare historical facts. They are also essentially dependent upon hypothetical and normative judgments. Besides a different evaluation of the present situation, an estimate of future development is also involved, and such an estimate is itself always based on certain assumptions. Above all, however, a judgment about the significance of modern technology for the fate of mankind decisively reflects one's definition of the essential, true nature of man: is man's nature fixed once and for all, or is it malleable, formed in each case as the historically contingent result of preceding developments? It is by no means an accident that those authors who have a critical or negative view of technology are informed, in their philosophical view of man, by the cultural and intellectual traditions of Europe, traditions which they see endangered by modern technology. Gehlen and Ribeiro, by contrast, with their optimistic views of technology, emphasize the anthropological development of mankind from its first prehistoric beginnings, while Lem and Moscovici go back even farther and integrate technological development with biological evolution.

It follows from what has been said that modern technology is not assigned an artificial, unnatural character merely for the reason that interventions in the natural processes occur, but because those interventions are undertaken in a specific manner and to an extent never before known. The critical step consists in replacing, on a large scale, 'organic' technology, which is based on the simplest tools of the artisan and on sources of energy provided by nature (wind, water, animal power), with mechanized 'inorganic' technology. Of crucial importance is the systematic use of continuously repeated monotonous rotating motions on which almost all mechanical processes are based (Diesel 5.3, 34). With the invention of the lathe with its tool-rest, allowing one to work metal in a symmetrical rotational manner, with high precision, a new realm of machine design was opened up, one no longer modelled after manual work, and even less trying to imitate nature (67–68). Yet it was not until mechanical power engines were developed (steam engines, later, internal combustion engines and electric motors) that the large-scale utilization of machine tools was possible, a development which opened the road to the Industrial Revolution.

Direct technological action embodied in artisan work is thus replaced by machine technology from which man is separated: manual utilization of tools, and human muscle power are replaced by technological systems and processes which in combination lead to desired results. While the utilization of wind, water and animals in organic technology remains subject to the rhythms of nature, and whereas artisan technology bears an understandable and immediate relation to the product, inorganic and mechanical technology is, from its very inception, based on uniform and anonymous processes. As Cassirer observes, this transition has as its consequence a loss of closeness to nature and of immediate significance: "In the moment in which man signed himself away to the harsh law of technological work, a multitude of immediate and spontaneous pleasures, which organic existence and purely organic activity had afforded him, were lost forever" (*1.9*, 47).

Regardless of whether one assesses this process negatively and deplores it as a dwindling of life's ultimate significance, or, on the other hand, hails it as liberation from the yoke of physical labor and a first step toward higher forms of human existence, the fact is that the loss of an intelligible and immediately significant lifestyle is inevitable if one is to claim the positive achievements of modern technology. Alienation caused by technology is the unavoidable price of physical convenience and of the increased potential for utilizing the immaterial world. In order to be able to actually utilize the increased efficiency of mechanical procedures, one must submit to their immanent logic.

This always involves, of course, a certain range of actions. Instead of rigorously subjecting the worker to the monotony of highly specialized procedures, it is possible, for example, to humanize human labor through the introduction of variety and change into the work assignment. The problem is one of balance between technological and economic efficiency, on the one hand, and the interests of persons directly involved in the technological procedure, on the other. Although progress in engineering opens new roads toward a more humane technology, such possibilities are, nevertheless, limited by the demand of the technological processes themselves. For example, the advantages of serial production can be realized only if a large number of uniform products are produced. Similarly, a continuously running technological process requires constant monitoring. Above all, however, the steadily increasing complexity of technological systems and processes would be unthinkable without a corresponding specialization and division of labor. Although it may not be possible to completely remove the undesirable consequences of severely limited, monotonous activity, one can mitigate them. What has been illustrated here through the example of mechanical processes, which are historically the earliest technological processes and possess fundamental theoretical significance, applies with appropriate modifications to all areas of technology.

In analogy to the biosphere, which is formed through the influence of microbes, plants, and animals on the inorganic world, one could term the technological environment created by man as the 'technosphere'. While in earlier times the technosphere was perfectly integrated into natural processes and the established balanced state of the biosphere, it has today become a force in its own right, increasingly determining conditions of the biosphere. The *dominating role* of the technosphere, with its specific mechanical and inorganic functional interrelationships, thus constitutes what is actually artificial and unnatural about technology. The dominance of the technosphere destroys not only the unquestioned integration of technological activity into the life of nature, with its varying rhythms, but also obscures the sensual perspicuity of the natural process. The objective approach to nature taken by modern scientists and engineers regards the physical world no longer as a self-evident and structured entity but rather as a mere conglomeration of matter that obeys abstract mathematical laws. By way of appropriate instruments and devices, technology provides the experimental basis for the 'de-sensualizing' of nature.

This technologically mediated, abstract concept of nature constitutes the starting point for the transformation of the given 'first' nature into the 'second' one of the technosphere. Today this technology-created environment totally surrounds us, and one may suspect, as Heisenberg has suggested, that for the future it will become as inseparably a part of us "as the snail's shell is to its occupant or as the web is to the spider" (*5.14*, 18).

Already today, the external dimension of life in the industrial nations is completely shaped by technology. Man has managed to control the threat posed by epidemics, famines, and natural forces with the help of technological aids. Since, however, the technology he confronts is of his own creation, what he is indirectly facing is actually himself. It is no longer a merciless nature which threatens him but rather the competing intentions of others as well as his own excessive demands. If one further considers the fact that the modern view of nature is based on technological experiments and hypothetico-deductive theories, the present situation can be summarized thus "that *for the first time in the course of history modern man on his earth now confronts himself alone*, and that he no longer has partners or opponents" (Heisenberg *5.14*, 23).

Nevertheless, the efficient functioning of complex technological systems must be bought at the price of a high degree of specialization. In contrast to the relatively unspecialized characters of artisan technology, the highly developed procedures of modern technology are based on a complex interaction of various subsystems, each designed for a specific task. A failure of any one of them usually leads to system-wide failure. Modern utility, transportation, and communication networks are particularly vulnerable to technical malfunction, natural disasters, or acts of sabotage. Whereas, in former times, mankind existed at the mercy of the natural environment, it is today no less dependent on the smooth functioning of the man-produced technosphere. If an atomic war should destroy this realm, the primitive tribes that are untouched by technological civilization would have the best chance for survival.

2. THE COSMIC DIMENSION

It is commonly believed that history and nature are exclusive opposites. Natural processes, which are not influenced by the human hand, recurring rhythmically, or remaining unchanged, stand in contrast to ever-changing, unique, and unrepeatable historical events, which are (intentionally) produced by man. 'Nature' may be defined here in a twofold sense: Firstly in the *material* sense as the sum total of the concretely shaped and sensually perceptible material world, which remains practically unchanged in its phenomenal quality during the span of a human lifetime. Secondly, in the *formal*, abstract sense it refers to the totality of natural laws to which the concrete phenomena are subject; it is usually assumed that the structural determination of natural processes is time-independent.

The picture is different if, following Moscovici, one focuses on the development of the universe (*1.24*, 43, 44). For in that case it becomes apparent that nature, like mankind, also has a history. From this cosmic perspective, the sharp contrast between nature and history is, in a sense, an untenable position, since both are subject to change in time. From this point of view, prehistory up to the onset of written records appears as a minute portion — and actual history as an even smaller fraction — of an extended process which led from the 'beginning' of the universe, through the formation of galaxies and planetary systems, to the birth of life on earth, to the subsequent evolution of disparate biological forms, and, finally, to the appearance of man. Thus, in the sense that both nature and history comprise a sequence of concrete consecutive changes, they are similar, as long as the question of causation, whether through spontaneous natural processes or by human actions, is not considered. (The important question as to whether the course of evolution is completely determined through the laws of nature, as opposed to the open historical process, cannot be raised at this point.)

The analogy between nature and history can be pushed even further if one considers the historical development of mankind in terms of technological activity. In this case, technology appears as the continuation of natural history, for it is by technology that the transformation of the world, begun in cosmic and biological evolution, is carried forth through conscious human shaping. The spontaneous and natural processes of evolution are now replaced by technological action, by means of which man — as Moscovici sees it — makes the history of matter *his* history. The technology of any given era is the result, Moscovici further claims, of purely intrinsic social processes which are independent of incidental particular goals and which occur according to the principle of natural selection and as adaptation to the changed conditions of life (*1.24*, 45—46). The historical process is based on social processes, and these, in turn, are determined by the requirements of social production and consumption; as a result, Moscovici considers "society as a form of nature (*la societé comme forme de la nature*)" (562).

This speculative interpretation of technology in the sense of an evolutionary theory of nature has — besides problematic details — to struggle with the difficulty connected with any monistic theory: since all relevant phenomena are to be reduced to one comprehensive principle, the respective key term necessarily becomes abstract and largely void of content. Thus 'nature' in Moscovici's theory, as the extra-human, material world, represents, on the one hand, the *object* of technological activity; on the other hand, however, human society, as the *subject* of this activity, is also considered to be a part of nature.

The mediation between the two aspects of nature is achieved by means of technology, thereby establishing again its double significance: by creating a technological 'supernature' through his work on the given natural environment, man distances and alienates himself from original nature to which he irrevocably belongs *qua* biological entity, and, in this very process, appropriates nature in modified and higher form for his own purposes. On the one hand, he sees himself in confrontation with nature as the object he must fight and dominate; on the other, he is, after all, one among many of nature's creatures. Such a contrariety, which can be analyzed in detail, is, nevertheless, rooted in the nature of things.

When considering the history of mankind from the point of view of struggle with nature, the transformation of the physical world effected by creative technological actions appears indeed as a mere microphenomenon within the development of the universe. Just as with any other interpretation of newly occurring phenomena within the continuing evolutionary process (*emergent evolution*), the problem is how to explain the origin of previously unknown forms of technology. If 'nature' refers not only to the material world to be acted upon, but also the ultimate basis and driving force of cosmic and biological evolution, *as well as* of historical development, then such a concept is overburdened. In the last analysis, the choice appears to be between a naturalistic and deterministic conception or an interpretation that accounts for the spontaneity of the historical process through recognition of a specifically intellectual dimension. However, neither the reduction of technology to a mere natural process of *practical* mastery of life in the context of cosmic and biological evolution, nor its derivation from purely theoretical, *intellectual* components is satisfactory since, in both cases one aspect is taken as absolute. As with man himself, so, too, the technology he has created belongs, in its physical concreteness, to the material world of natural processes while, at the same time, being the result of creative intellectual and cultural activity, it reflects the spontaneous, conscious, and purposeful transformation of nature.

Lem, like Moscovici, regards technology as an extension of biological evolution. There are, he points out, numerous analogies between "bio-evolution" and "techno-evolution": both are based on *mutations*, i.e., the rise of new forms among which, in the struggle for survival, a *selection* takes place, with only the most fit species surviving. Furthermore, in both cases the first forms are extremely primitive. Only with difficulty can they prevail against the existing species which are already better adapted to a given task. Through shifts in the environment, an explosive expansion occurs, taking a large variety of forms, and leading to a climax of the development: "In its 'struggle for survival' the automobile not only displaced the stagecoach, but also gave birth to the bus, truck, bulldozer ... and dozens of other vehicles" (*5.23*, 28). When a disturbance in the homeostatic equilibrium occurs, gigantic forms may emerge as the system reacts to preserve an existing species, a process which frequently leads to overspecialization or the survival in ecological niches of evolutionary relics. Techno-evolution and bio-evolution can both be regarded as self-organizing systems with "internal" programs, capable of utilizing anything in the universe for chosing structures or building materials (24—27).

Lem holds that techno-evolution, like bio-evolution, is an organic process outside any moral laws. Just as there is ultimately nothing permanent in the physical world, so, too, in the realm of history, systems of morality are subject to change, a fact made apparent when one surveys long stretches of time and in particular prehistoric evidence. Lem considers it a fundamental error to regard technology as the creation of means and tools which are then judged and utilized according to moral canons. As he sees it, the opposite is actually true. Technology itself forms our moral attitudes. In Lem's opinion, past and future development of technology is not determined by conscious choice and planning but rather through intrinsic dynamics which lead to the realization of all potential technological actions (5.23, 52–64). In this process, man cannot remain the last "authentic work of nature." The level of achievement of medical-technological information makes mandatory the systematic reconstruction of the species by means of a breeding process incorporating genetic manipulation. In this way evolution, which has up to this time occured spontaneously becomes consciously directed, and "the next model of *homo sapiens*" must be designed for new and biologically higher values (501–503).

Informed by biological evolution and formulated in the style of science fiction, these ideas, with their icy logic, open up frightening perspectives. They are the predictable expression of a style of thinking entirely directed towards what is scientifically and technologically feasible. The book title *Summa Technologiae*, chosen in analogy to the dominican Thomas Aquinas' well-known summary of medieval philosophy and theology, suggests a fundamental shift of perspective. The spirit of the times is no longer expressed in a philosophically elaborated theology, but rather in the idea of technological progress. For Lem, religious ideas are themselves simply the result of spontaneous evolutionary processes aimed at "optimal adaptation to conditions entirely of earthly existence" (5.23, 211).

One may discount his demand for "replacing anthropocentrism with galactocentrism" (130) as being a distortion of the time perspective as well as an unrealistic misconception of the concrete conditions of man's existence; nevertheless, Lem's theory of similarity between biological and technological evolution has a definite superficial plausibility. The suggestive force of his hypothesis, due in part to his fantastic manipulation of the time dimension, stems mainly from a mixture of quite appropriate *descriptive* elements and extremely problematic *normative* ones. The statement that those technological innovations become generally accepted which turn out best in the "struggle for survival" is certainly correct (tautologically). Furthermore, no one can be blind to the fact that, in recent times, technology has been steadily growing, largely independently of individual intentions and in a quasi-organic way. Nor can one fail to recognize the tendency toward actual realization of all new action possibilities thus created.

On the other hand, a *description* of present conditions should not be confused with an *evaluation*. It is one thing to acknowledge the fact that technology is transforming the world on a large scale. It is quite another thing to approve each instance of this process, and to want to intensify it evermore in the future, going even further to apply it to all areas including man. At this point Lem is operating with the general premise of a *technological imperative* according to which all possibilities for action must be realized simply because they exist. He considers as positive and progressive that new breed of man that proves superior in the struggle for existence, justified by the 'right of the stronger'. Lem thus completely ignores the intellectual and cultural aspect of man which elevates him above mere physical existence, by reducing this aspect to biological functions or the accumulation of empirical facts. Properly speaking, this position thus turns out to be inhuman, since its systematic realization would sacrifice to the 'law of the jungle' all intellectual and cultural achievements of mankind that transcend animal existence, as well as the freedom and basic human rights of the individual.

Lem's theorizing constitutes a challenge. By its radicality, it points up the inescapable consequences of technology left completely to itself; in this way it can contribute to a reorientation and conscious self-limitation. At the same time, however, we can take the fascination exerted by his fantastic visions on a large audience as an indication of how common it has become today to think in categories of technological feasibility — categories which originated in modern (European) history. The application of medical-technological knowledge on man through organ transplants, artificial fertilization, and possibly also genetic manipulation can be taken, in this respect, as a sensitive indicator.

It is true that, up to the present, individual rights over one's personality and body have generally been respected. What is, however, a cause for grave concern about Lem's proposals, is the fact that they explicitly legitimate man, in his biological and physical constitution, as a field of experimentation in which to 'try out' every conceivable modification and mutation. With this, the objectification of nature, one of the prerequisites of modern technological action, is applied to man himself. Now he, too, is to be considered like any other material object, a thing to be arbitrarily dealt with. The last bastion of resistance to technology has now been breached.

In the *Foundations of the Metaphysics of Morals*, Kant ascribed to the autonomous, rational personality an inalienable value, dignity, and the status of an end-in-itself which preclude mere instrumentality. "But man is not a thing — not something to be used *merely* as a means: he must always in all his actions be regarded as an end in himself" (*4.14*b, A 67). Practical reason, which for Kant is based on the autonomy of the person and is aimed at the establishment of a moral world order, is for Lem nothing more than an instrument for collecting information, i.e., positive knowledge, and for providing instruments with which to expand the range of technological feasibility as far as possible. While the moral inhibitions that bar the use of man as an experimental object still stand today, the question is when will they break down?

Technological action is based on intellectually conceived, goal-directed transformation of the physical world. Technology is thus a part of both the intellectual and material spheres. Accordingly, speculation about technology can take two directions. One is to emphasize the material component of technology, as Moscovici and Lem have done, and thus regard technological activity as a broadly conceived natural process. In this case, technology is merely one element in the physical development of an entirely material universe. Alternatively, to emphasize instead the intellectual component is to reverse the cosmic perspective by interpreting technologization as the spiritualization of matter. In terms of the framework of a philosophy of technology, this view also reflects the traditional opposition between matter an mind.

Along these lines, Werner G. Haverbeck speaks of technology as leading to a "humanization of nature", since, by his actions, man conceptually grasps processes begun in evolution — flying, for example — and furthers them by technological means, thus revealing "nature's dependency on man". With the possibilities supplied by technology, man can penetrate all areas of the physical world and can grasp them intellectually, becoming thus the "consciousness of all-encompassing nature". It is in this way that Haverbeck sees the "purpose of technology" as the "humanization of the earth" (*1.13*, 273—78). The theological interpretation of Beck also reflects this point. "As nature enters more and more into the *creative* spirit of man, man himself enters more and more into the creative spirit of God" (*1.6*, 64).

Similarly, Andreas van Melsen sees man as "mind-in-matter" and thus considers technology as an indispensable element of all intellectual culture. Once intellect completely understands itself and its potential, recognizing nature as matter to be worked, it cannot but realize the given potential for technological action: "To become human does not mean only to form and enrich the spirit, but also to work nature so as to put it under the control

THE TECHNOLOGICAL WORLD 133

of the spirit, because it is in this control that lies the destiny of matter" (*1.21*, 256).

It is noteworthy that these 'intellect-oriented' interpretations of technology lead in principle to the same positive evaluation of technology as do the 'matter-oriented' ones, even though they are based on different metaphysical presuppositions. In contrast to Moscovici and Lem however, Haverbeck, Beck and van Melsen are keenly aware of the dangers of excessive technological development. An unprejudicial assessment of technological development, however, compels one to question whether a monistic approach — even in the form of universal metaphysical speculations — is at all capable of describing the phenomenon of technology. For *prima facie*, neither the arguments of materialism nor idealism are refutable; man is an organic material being, and the spread of modern technology on the whole resembles a natural vegetative process. At the same time, however, man is a thinking being, conscious of himself, and his technology reflects in its details theoretical, purposive, and planned activity. In this situation it is probably appropriate to resolve the apparent contradiction with a complementary dualism.

3. ACCUMULATION AND 'SELF-REINFORCEMENT'

Historical phenomena such as the expansion of cultures, or the formation of political domains are subject to the rhythms of growth and decay. After an initial growth phase, a climax is reached, to be followed by a stage of decline and reduced development, or a new beginning. The historical record abounds with the greatest variety of advanced cultures and large empires, each of which blossomed for a time. Such unique processes, which are comparable to the growth of a plant, are based on a temporally contingent coincidence of appropriate prerequisites. The processes stagnate, or cease to exist once these conditions are no longer present. In striking contrast to all other historical developments, modern technology has, up to now, enjoyed continued quantitative and qualitative growth. This growth has included steady progress within the engineering sciences, the increasing expansion of technology into more and more areas of daily life, and the geographical spread of technological procedures and artifacts across the world. This almost irresistible and ever accelerating growth has become *the* determining force for societies, cultures, economies, and political systems, shaping the face of our time and pushing us toward a uniform technological world civilization.

The phenomenon of steady growth of modern technology — unique in world history — must stem from a very special set of causes. Since technology is spreading these days independently, for the most part, of cultural and social preconditions, one might first try to locate the real causal factors within the structure of technological activity itself. True, technological artifacts themselves, representing the state of the art, constitute a stabilizing element within such a process. The actual physical existence of technological systems, however, can at best explain the continued existence of present technology, but could never explain the dynamics of actual technological development. It is therefore obvious that there must be additional reasons for the expansion of modern technology.

Furthermore, it is self-evident that technological development *as a whole* cannot be attributed to conscious decisions and goal-oriented acts of individual human beings, but like other comprehensive social and cultural phenomena is accomplished as the result of various, possibly even contrary, intentions. One can even observe here a *paradox* of technological action: construction and production of technological objects is based on carefully planned and purposefully executed actions. Procedures of the engineering sciences are thus rightly considered models of carefully designed and rational action. One should expect accordingly that technological development is in every respect subject to human goal-setting and control. The well-contemplated and logical action to which technology owes its efficiency is, as a rule, restricted to only the immediate *technological task*.

Beyond this, all technological activity is always embedded in a specific *social context*. Production, distribution, and utilization of technological products are part of a very complex process characterized by a division of labor, involving, in one way or another, all members of society. Therefore, the concrete consequences of technological actions go far beyond the mere solving of engineering problems. Furthermore, they are by no means limited to the direct handling of products either. Thus, work roles, which have become depersonalized and streamlined for smooth functioning, are just as much influenced by technology as are standardized habits of consumption and leisure as well as the intellectual self-definition of the times. Technological progress, which is deliberately chosen, well-conceived, and intentionally created at the level of detail, has thus, as a whole, assumed the character of an independent and hardly controllable historical force determining the fate of mankind. This paradoxical phenomenon of a self-created technology which is at the same time an elemental alien force, provides the basis for interpretations of the global development of technology in both idealistic and materialistic terms. Modern technology must therefore possess a built-in mechanism for accumulation and reinforcement whereby individual measures of engineering technology lead, in their totality, to a continuous, and apparently irresistible, process of technologization. It is this intrinsic dynamic that explains the almost explosive growth of technology in the two hundred years which began with the Industrial Revolution. More specifically, the following four factors can be distinguished:

(1) In contrast to other historical processes, one can observe, in the course of technological development, objective progress which

produces ever increasing yields for the same amount of manpower and material resources. Since man is always dependent on a certain form of technology, more perfect and efficient solutions satisfy a potentially basic human need, one which nevertheless was not actualized as a generalized behavioral norm until the late Middle Ages or even not until the Industrial Revolution. Since, in the last analysis, technological systems always yield tangible results, states, or starting conditions, their degree of efficiency is, as a rule, also immediately evident. Of two competing conceptions the one most favorable to the relevant criteria (efficiency level, operational safety, production cost) will always be the one realized. On the basis of this *selection mechanism*, only those variants will succeed that amount to actual improvements over the present state of affairs. That this is true becomes especially clear in a comparison of somewhat longer time span, e.g., compare the present state of transportation and communication technology with that of just a century ago.

(2) In addition, there is the *accumulation* of technological achievements. A newly acquired insight, procedure, or device enlarges the repertoire of what has already been achieved and thus becomes an integral part of technological knowledge and ability on which to relie for any future innovations. While theoretical knowledge and practical skills must be imparted by instruction and training, the material state of development is, without any intermediate link, physically 'preserved' in the form of the latest devices and instruments. In most cases the entire stock of preceding achievements can be brought to bear on a new problem. It is obvious that through this process of accumulation, steady technological growth is likely, with technology transfer contributing to an overall expansion.

(3) The principles of selection and accumulation are intrinsic to technological action and therefore applicable to all eras in the history of technology. Because of them technology, independently

of the socio-cultural context, has generally shown steady development. The history of technology may, therefore, be described as a continuous process determined only by technology-immanent, or 'internal', relationships and not by changing cultural, social, and economic conditions. By themselves, however, these principles cannot explain the expansion *by leaps and bounds* of technology during the last two hundred years; for this, additional determinants must have been decisive.

One of the two specific factors on which the upsurge of modern technology since the Industrial Revolution is based, is its *systemic character*: the technological innovations with which the process of industrialization began were not isolated phenomena. From the very beginning they were systematically related to each other in many reciprocal and complementary ways. Thus improvements in iron production and steel processing (blast furnaces, rolling mills, machine tools) supplied the necessary prerequisites for bridge construction and high-rise buildings, and at the same time made possible construction of power engines (steam, electric, internal combustion) by means of which new energy sources (coal, oil) could be utilized. These in turn gave rise to new machines and procedures in the textile industry and produced improvements in transportation (railroad, steamship). This close interrelationship between various technological subsystems is far from accidental. It is the necessary result of the basic objective of modern technological activity — the exhaustive and efficient utilization of every technological possibility. In this way different types of technology have developed in a relationship of mutual complementarity. As a result, the combination of materials, construction elements, and procedures gathered from many fields always leads to new possibilities for technological action, which, for the most part, are then also realized. Thus, with regard to the level of technological achievement existing at any particular time, every fundamental innovation has a multiplier effect on the overall process of technological development. The higher the general level of technology itself, the greater the result produced by this effect. Of course, the systemic character, i.e., the fact that different subsystems are objectively interrelated, thus leads to a systematic self-amplification of modern technology, since every innovation leads simultaneously to both the *directly* intended effect and, as explained, an *indirect* one.

By a succession of detailed improvements, each procedure moves along to the point of actual application. Since for practical research and development the entire range of existing technology is accessible, one is assured of at least partial success. In this way, entirely new branches of industry can emerge, e.g., computer or nuclear technology, and themselves in turn stimulate further innovations in other fields. This accumulative process is further enhanced because the many problems, themselves created by technology, require even more technology for their own resolution. For example, problems of preservation of the natural environment, disposal of radioactive 'waste', potential misuse of computerized information storage, are themselves the side-effects of certain technological procedures. A radical remedy would be to eliminate such technologies entirely, thus doing away with the unwanted consequences. As a rule, however, no one wants to do this, so we are left with the choice of coping with the problems by means of appropriate counter-measures such as, for example, purification or filtration plants. While this does not necessarily lead to a spectacular growth in the quality of technology, it does suggest that the negative material side-effects of new technologies can only be contained by means of 'more' technology. In the case of organic craft technology with its comparatively minor alterations of the biosphere, such problems did not arise, or else they were automatically corrected by nature's regenerative capacity. By contrast, the complex artifacts and large-scale interventions into nature by science-based inorganic technology lead necessarily to further expansion of technology, whenever existing technological potentials are fully exploited.

(4) This intrinsic mechanism of expansion requires the continous availability of new and refined technological procedures. Such innovations, however, do not just happen. They presuppose the existence of applied science and empirical investigations. Thus, the second prerequisite for the rise of modern technology can be seen to be the appropriate *research methodology*, which contains as a decisive element a mutual complementarity between the sciences and technology. While it is true that this complementarity did not become effective on a large scale until the second half of the 19th century, it has since that time increasingly determined technological development. Nevertherless, conceptually at least, the complementarity between modern science and technology had already existed in terms of their objectives: both activities are concerned with isolating certain natural processes with the aid of appropriate procedures, with the only difference being a question of goals, one aiming at immediate practical utilization, the other at detached theoretical understanding.

Therefore, on the one hand, scientific research always makes technological application possible; on the other, technological devices and apparatuses open up new opportunities for scientific investigation. An advance in the technological 'state of the art' leads to new scientific instruments, while producing new knowledge in the process. This, in turn creates new potential applications for technology. If one further considers the fact that technologically generated problems raise questions which demand scientific answers, then one realizes that the close interconnectedness of the two fields obscures any clear line of demarcation between them. This interplay of basic and applied research, new to the history of technology, enlarges the realm of technological potential.

There exists in the methodological approach itself, with its theory-guided, systematic utilization of natural forces, an intrinsic tendency towards technological self-reinforcement. During the first phase of industrialization, which was primarily characterized by artisan technology, this feature was not prominent. Nevertheless, the present state of technology would clearly be inconceivable without the results of scientific research and the introduction of the methods of engineering science. In this regard, Moscovici sees the decisive feature of modern technology in its general creativity, manifested in a process of continuing technological refinement. For him, domination over nature does not consist primarily in the manipulation of material objects and the application of existing skills, but rather in the creative institutionalization of the work process ("*la création du travail*"), with man using his powers and abilities to acquire ever new talents, skills, and knowledge beyond the existing state (*1.24*, 47–48). In terms of this view, therefore, *concrete action processes* and the *physical transformation* of nature are completely subordinated to the *methodological dynamics* of the development of technology. Despite its one-sidedness this view addresses a basic characteristic of modern technology.

4. THE ACTING INDIVIDUALS

In the above discussion the process of technologization was frequently, by way of abbreviation, represented as an independently occurring process. Such formulations serve a purpose and are even indispensable, if complex matters are to be expressed in a simple and clear manner. Any abbreviated description, however, even though it possesses practical utility, will by itself supply a fundamentally wrong picture by elevating technology to the rank of autonomous subject. The categorical error on which such a hypostatization is based has grave consequences because of the impression created that technology runs its course without human involvement, subject instead to some inexorable natural law. Ellul, for example, continually presents technology as an autonomously

THE TECHNOLOGICAL WORLD 141

acting subject (5.4). It is only logical then, that his conclusions are overwhelmingly negative and void of alternatives, since man seems to be faced, without help or recourse, with the demon technology which devours everything, usurping and exploiting it. Similarly, in orthodox Marxist literature, technological development is frequently described as an inexorable process, since conceptual constructs are treated as real entities.

Whenever the determinants of the technological process are being discussed, however, a proper analysis requires that the concrete action processes to which technological systems and processes owe their existence be accounted for. In particular, the trivial point must be made that technology, like science or culture, for example, does not exist by itself. Everything that exists in the realm of technology has been created by man and is therefore dependent on his dominant values and goals at any given time. Oversimplified formulations, in which it is said, for example, that technological development leads to certain results, or that present levels of technology dictate that certain steps be taken, must, in the final analysis, always be related to human acts and intentions. In this way it is made clear that we are not dealing with completely irresistible and necessary processes, but rather with phenomena similar to other social and historical developments, all of which are capable of unpredictable shifts and deviations from past trends.

This fundamental claim does not contradict, but rather complements, the previously discussed mechanism of accumulation and self-reinforcement. Principles of selection and accumulation are no more independent and necessary determinants, than are the systematic character of modern technology and the methodological relation of complementarity existing between technology and the natural sciences. In particular, the expression 'self-reinforcement of modern technology' needs to be understood metaphorically, since strictly speaking what is referred to are certain relational properties which are characteristic of a technological approach aimed at an increased domination of nature, maximum efficiency, and the systematic exhaustion of all possibilities. Methodologically this produces only *hypothetical* instructions for action: statements of how to proceed *if* a certain kind of technology is in fact desired. These statements are binding only under the assumption that the 'supply' of action possibilities resulting from the existing state of technological development is actually to be utilized in some specific sense.

Precisely because it is human beings who realize any given technological potential, guided in their actions by conscious intentions or unconscious motives, an appropriate change of human values will cause a change in attitudes toward technology. In any particular case, however, there are certain limits that must be considered. Thus, the existing state of the art of technology presents a *physical* limit which weighs more heavily the higher the degree of technological development is. Because of the existing technological systems and the necessary support devices there is always present a momentum for the continuation of preceding practice, since radical change would necessarily lower the level of output. Added is the fact that with rising technological potential, the level of demand also rises, so that previous developments create a kind of *psychological* inertia. On the whole, however, this much remains true, the autonomy of technology extends, in principle, only as far as the acting subjects accept the established trend and as long as a general optimism toward such trends continues to prevail.

This does not, of course, preclude the possibility that behind the backs of the individuals there is a long-term and universal technological process at work, for which the individual is merely, as it were, the executor. Rather than existing by itself, such a process, to become concrete, requires individuals who actually conform to the 'demands' of technology. While it is true that only within narrow limits can the individual free himself from generally accepted opinions, views and attitudes, such concrete manifestations of the spirit of the times are never autonomous entities. Rather, they are themselves mediated by tradition and subject to historical change, always susceptible to human approval, modification or rejection. Were this element of spontaneity and freedom lacking, man would be completely at the mercy of the laws of nature, totally void of choice or selection as to the inheritance passed on to him. Despite the inertial force of historical precedent, there is always the theoretical possibility of a 'quest for new shores' as witnessed by the unexpected turns in previous historical developments, and as further supported by the impossibility of making any reliable predictions about future technological developments.

Because modern technology is the result of a complex process, based on the devision of labor and almost incomprehensible, it would be hopeless to try to exhaustively specify all the relevant evaluation and decision processes that influence the actual course of technological development. Complicating the matter even further is the fact that the acting subject must always take into account the actions and reactions of his fellow human beings, not to mention the unforeseen consequences of his own acts. Nevertheless, this is in principle the starting point for a non-fatalistic analysis of the development of technology. A grave impediment to such an investigation, however, is the fact that human choice is, as a rule, by no means based on consciously articulated decision procedures, but rather is largely determined by unreflected internalized behavioral patterns. Nevertheless, the hypothetical reconstruction of actual decisions – e.g., according to a model of goal-oriented, intrinsically rational selection behavior – is the only alternative to a mere theory-free description of the facts.

5. INDIVIDUAL FREEDOM AND COLLECTIVE TASKS

The general theoretical statement that in the present situation, just as in all other stages of history, the future is in principle open, can hardly be disputed. Yet, just as incontestable is the empirically observable fact that as technological developments increase, the range for alternative options decreases proportionately. Subjectively, this situation is today perceived as oppressive or even as an immutable fate imposed on all mankind. In this connection, one can observe a basic ambivalence in attitudes toward technology: people regret that life has become increasingly plannable and predictable, and yet, at the same time, long for this situation.

Different reasons account for this ambivalence. Reliable prediction or complete calculability of the future produces a paralyzing effect and leads to fatalistic passivity, since there are no more true challenges to be mastered through special personal effort. The possibilities for spontaneous self-realization, individual development, and for proving oneself become of necessity more and more limited in a standardized technological world, designed in all areas of life for smooth functioning, as technology develops and its procedures acquire new levels of perfection and efficiency.

In principle — if not in all details — technology-related alienation is the necessary and unavoidable consequence of the advantages granted by technology. Relief from tiring physical work, safety from the inclemencies of nature, and the amenities of technological luxury can only be had if one is willing, at the same time, to accept the intrinsic logic of technology. Metaphorically speaking, we have created 'a second nature' by means of technology, and it surrounds us like a spider's web. It protects us against unforeseeable influences and offers an enormous variety of conveniences and reliefs. We should, therefore, not complain upon finding ourselves indeed ensnared upon all sides by this web.

Today, however, it is not just technology-related limitations on personal freedom and life-style options that prompts complaints. *At the same time* there is a cry for even higher measures of control, planning, and foreseeability. This is not without reason. All technological achievements are, after all, based on well thought-out, planned and systematically executed procedures. If one were to proceed spontaneously, arbitrarily, and without forethought the technological system would inevitably break down. The ability to fully utilize the advantages of technology requires the most extensive planning and control. As the technological process intensifies, the mechanisms of accumulation and self-reinforcement, reflecting the inner logic of technological action, become more and more effective. Man himself is firmly integrated in these self-created artifacts and their functional interrelations, since technology serves man only to the extent that man serves technology. Here again the ambivalent character of modern technology is revealed: on the one hand, it offers relief from the servitude of labor, thus creating, at least in principle, the opportunity for a higher form of human existence. Yet this same technology inescapably creates rigid and impersonal life-styles and thereby many different forms of alienation, which are themselves contrary to the ideal of the free development of the individual.

Since this contradiction between spontaneous self-determination and long-range planning of living conditions is an essential component of modern technology, its effects can never be completely neutralized but only mitigated. The ideal of the *autonomous individual* as the highest goal of human existence is part and parcel of the European philosophical tradition traceable to antiquity and elaborated by Rationalism and the Enlightenment. Work in industrial technology, however, which is likewise based on Western traditions, with its method of mathematical-experimental domination of nature, depends, in contrast, on carefully designed and systematically executed *collective actions*. Technology as we know it today, is the result of processes of social action in which virtually all members of society participate. The realization of its aim of maximum efficiency requires a maximum of coordination and standardization. This trend toward uniform planning and control is further reinforced by the needs of social security and the many social services, all of which assign to government authorities a host of guidance and control functions.

Thus the dilemma becomes obvious. Technology is supposed to serve man, and, in particular, to provide the conditions for personal development. This requires that there be available a maximum range of individual freedom. On the other hand, however, technological functions require planning, coordination, and controlled action, and this necessarily limits the range of individual decision-making. In the limit, individual freedom of choice and self-determination is confronted with general control and planning which are required according to the criteria of technological efficiency.

Further exacerbating this situation is the fact that the actual utilization of a particular technological potential often reflects the limited horizons of special interest groups, while the actual consequences fall on a much larger group. As such, this is not unusual, since one can point out for any action — if the context is sufficiently broadly defined — direct or indirect consequences which go beyond the scope of original intentions. In the case of 'organic' technology, however, based as it was on relatively isolated and minor interference with the natural process, such extended side-effects were narrowly limited, and could therefore be neglected. In contrast, the large-scale systematic exploitation of nature, characteristic of science-based industrial technology, and the added fact that various technologies and processes are increasingly intertwined, make a consideration of higher-level criteria imperative.

In this connection Emmanuel Mesthene offers a fundamental critique. He observes that technological innovations initially reflect the descisions of a handful of individuals using criteria such as increased productivity, better sales, or higher efficiency. Concerns about social advantages or disadvantages are not decisive. The aspects that are relevant for the entire society amount to side-effects, at first not even envisaged, but nevertheless bound to occur. The mistake, resulting from a lack of appropriate institutions and organizations capable of dealing with unwanted side-effects, consists, according to Mesthene, in not considering from the very start 'social costs' (such as pollution) and of technological projects. What is responsible for the undesirable effects of technology is, in his view, the fact that decisions are influenced by individual and economic considerations rather than determined on the basis of social obligations. To remedy this situation, he suggests that a social cost-benefit analysis precede all technological innovations based on the use of social indicators to measure overall societal impacts. This point of view assumes that there exists an objectively unavoidable competitive relationship between preservation of individual freedom and control of technology requiring social action for its maintenance (5.29, 37—42).

Thus, it turns out that modern technology invites planning and control for several reasons. One set of reasons is of an intrinsically *technological* nature, because efficiency and capacities of the increasingly complex technological systems can be achieved only through a maximum of planned and systematically executed measures. Secondly, however, the hyperactive development of technology makes *socially* mandatory certain counter-measures and control mechanisms to forestall the unlimited growth of unwanted technological side-effects. Finally must be noted the *psychological* willingness to accept planning and control which has emerged along with the process of technologization: as technology has become the determining force in all areas of life, it has come to serve as a model of orientation for understanding social and historical processes, to the end that social conditions and even the entire history are themselves viewed in terms of technological feasibility. In line with this technology-oriented mode of thinking, planning and feasibility are then even applied to man himself. Just as the concrete material problems raised by technology frequently can be coped with only by intrinsically technological means, i.e., through additional but different technologies, one is also forced in the case of social planning and control to continue along the road initially taken in order to exert a desired level of control over all developments.

Technology and planning, however, can also contradict each other. This situation is addressed by the technocracy issue, which is concerned with political domination in the modern technological state, itself subject only to technological 'inner necessities' and dependent upon the technical competency of experts (cf. Lenk 5.25). This discussion, however, presupposes the tacit assumption that in every case a quantitative increase of technological output is to be aimed at. In this case, planning loses its status as an instrument for solving problems which are freely chosen and extrinsic to technology itself, and becomes rather a tool of technological objectives. Upon closer scrutiny, however, the lack of choice possibilities proves to be illusory because the general premise of increased efficiency turns out to be the freely chosen goal. Everything else follows from this.

Even if the task at hand is to provide, in a rationalized and technology-oriented world, a larger measure of freedom and of opportunities for individual development, one cannot do without technological aids, plans, and implementation strategies. Speculative futuristic projections such as the utopian literature, beginning with Bacon's *New Atlantis*, or the 'dystopian' predictions of Huxley and Orwell rely on the role technology will play in the systematic and planned transformation of human life. It is noteworthy in this regard that until the end of the 19th century, technology was generally evaluated positively, because it was seen as the prerequisite for a general increase in well-being. Since then a fundamental change in attitude has occurred so that today there predominates in the futuristic literature a pronounced fear of uncontrollable and apocalyptic development in which man is no longer the *subject* of technology but rather its arbitrarily manipulable *object*. (For an overview of technological utopias, cf. Fritz Winterling 5.39).

Upon critical assessment, fundamental objections against the notion of total 'control' of human affairs need indeed to be voiced. The physical circumstances of life can admittedly be modified in goal-oriented ways through appropriate technological measures; the influences of technology in this regard are clearly undeniable. However, the most broadly conceived social structures and cultural values, which constitute the basic foundation of human existence, resist, by their very nature, all attempts to turn them into objects or to otherwise manipulate them. It is unrealistic to think otherwise, since man, without inherited values, would feel like a being in a void. This is the reason why all theoretically conceived changes involving plans and systematic implementation strategies are actually always subject to certain limitations. Thus, for example, new institutions and organizational forms can only fulfill their purpose if, first of all, they are built on traditional values and if, further, the persons involved actually accept them and make them work. While the physical world is arbitrarily manipulable, within natural limits, the social and individual situation is to a large extent inaccessible to planned and methodically executed control (cf. Wilhelm R. Glaser 5.10, 191 and Freyer 5.5, 84).

Utopias or planned designs which do not take this situation into account are therefore doomed to failure from the very outset. One can try, however, in analogy to the structural laws of the material world, to consider the objectively given conditions in this area in order to achieve what is possible in each situation. Concretely speaking, the following three aspects are particularly important:

(1) In the technological mass society, with its numerous political and economic interdependencies, there are no clearly distinct and functioning small groups capable of articulating a common will – along the lines perhaps of a democratic people's council – on

which pertinent decisions could be based. Quite to the contrary, the efficiency of technological procedures that is only attainable through standardization, together with the improved possibilities of transportation and communication, leads to comprehensive units which finally favor the emergence of a global system. Thus, the *delegation* of planning competency and political responsibility to the bearers of the legitimized democratic mandate becomes inevitable.

(2) Since man is not only rational, but, in his practical life, also a feeling and wishing being, it would indeed be an illusion to desire the creation of a generally binding consensus, solely by means of theoretical reflection on common goals. Due to differing interests among the various members of society, some form of *balance of interests* is always necessary. In a pluralistic liberal and democratic context this is only to be attained through compromise, with appropriate protection for minorities.

(3) The productivity of modern technology is based on the principles of specialization and division of labor, in particular. The complex details of any specific area are frequently intelligible only to the *expert*. Of necessity this raises the possibility of the misuse of power, since experts may give a one-sided analysis of any particular issue — a possibility somewhat offset by the fact that experts often disagree among themselves.

Because of the aforementioned difficulties, there does not seem to exist a procedure for the governance and control of technological development that is completely satisfactory. Nevertheless, one cannot disregard the necessity of achieving some form of long-range limitation on population, resource depletion, or nuclear armaments if a catastrophe is to be avoided. Considering the urgency of the situation, it makes some sense to envisage, as a thoroughgoing solution, some comprehensive system of planning and control. Such a concept does correspond to the basic approach of technological thinking which itself aims at maximum efficiency and thereby global control, standardization, and the elimination of points of friction. However tempting such a thorough solution of all existing problems may appear, it nevertheless proves, upon closer inspection, to be rather problematic, since any measures decided upon could only be implemented by a perfect power machine utilizing force and terror. In the last analysis, man would be degraded to the status of mere object, thereby losing everything that makes life worth living.

In this context Wolfgang Harich's proposal for rigidly disciplined and ascetically controlled technology is of interest (5.13). Harich takes seriously the theses of the book *Limits to Growth* (5.28) and argues for rigorous limitation of technological growth, a goal which he believes is only attainable through a communistic government with dictatorial powers. It is very noteworthy, however, that his ideas are not echoed by the Communist Bloc. This illustrates the fact that Eastern and Western attitudes toward technology, despite their economic and political differences, are much closer than superficial inspection would indicate. Technological progress and increase in living standards are sacrosanct ideals in both regions. The appeals of the critics of technology in the West have, so far, been just as futile as Harich's attack on the ideology of consumerism. The long-term damage caused by excessive technologization in both East and West is, in principle, attributable to the same pattern of behavior: the pursuit of short-term goals and the uncertain hope that, just as in the past, the future will yield (technological) solutions to all existing problems.

It would, nevertheless, be an exaggeration to see here only two alternatives for mankind: extinction as a consequence of a pathological growth of technology or a centrally distributed and apportioned form of global dictatorship. Just as even a few decades ago no one could reliably predict the present, so, too, future developments can only be probabilistically estimated. Furthermore it has been shown that in times of general emergency — e.g., wars or natural catastrophes — people are quite prepared to accept restrictions on individual freedom and on customary living standards, provided they see the necessity for such measures. Yet, considering that there are such different situations in different countries, it is difficult to imagine how mankind as a whole would submit to specific restrictions out of a concern for some future dangers, not presently apparent. A solution cannot be expected on the basis of organizational or institutional measures alone. What is actually required is a basic reorientation of values, and the real problem, given the ever-accelerating pace of technological development, is that such a change in attitude must occur very soon.

6. THE UNIVERSALITY OF MODERN TECHNOLOGY

The most obvious characteristics of our time is the ubiquity of technology. Wherever we turn we are surrounded by technological systems and processes which — especially in the industrialized nations — have fundamentally changed outward appearances and lifestyles. Geographically, too, technology expands more and more since nowhere in the world is there any serious resistance to technologization. This universal character of technology, as reflected, among other things, in the phrase "the age of technology" manifests itself in three different aspects:

a. The Transformation of the Physical World

The visible characteristic is the appropriation of nature for human purposes through a corresponding transformation of the physical world. Just as is the case with other characteristics of modern technology, the really noteworthy thing is not the bare fact of changing the environment, but the extent and effect of the technological actions. In order to assure his physical survival, especially in the face of a hostile nature, man has always depended on technology.

Up to the beginning of industrialization, however, such interferences with natural processes kept within comparatively narrow limits. As a consequence, the visible effects of technology were minor compared to the organic phenomena in the environment.

Today's world is determined by the greatest variety of artifacts, without which modern man could no longer even survive. This transformation is most visible in building constructions of all kinds (factories, streets, bridges, airports, residential areas), which, particularly in areas of high population density, determine the outward appearance. Together with industrial technology, these testaments to contemporary lifestyles continue to spread irresistibly across the entire earth. Untouched landscapes and rare animal species will in the future be found only in artificially protected reservations which are meant to limit technology's grasp. Another characteristic of the appropriation of the physical world by man is the utilization of raw materials and fossil energy for the fabrication and powering of technological implements everywhere. Furthermore, in all technological processes unwanted side-effects occur such as pollution of the natural environment by harmful waste products. The imbalances in the ecological equilibrium caused by this have assumed an extent which, in the long run, must call into question the earth's inhabitability, if present trends should continue.

The alteration of the material environment through technology is not limited to large-scale phenomena alone. The universal character of modern technology means that there is practically no area of physical activity remaining that is not, in one way or another, affected by technological artifacts or processes. Even in such fields as medicine or agriculture, in which organic processes dominate and which should therefore hardly be accessible to technology, technological devices, involving inorganic and mechanical processes, play an increasingly important role. Only in rare and exceptional cases − for instance, for purposes of leisure and recreation

— is modern man still dealing with nature left to itself and not altered by technology. As a rule modern man's physical environment consists of a wide range of apparatuses and devices, each serving a particular technological function and thus mediating between man and untouched nature. This is particularly obvious in the case of communications and transportation technology, where technological constructions of the past two hundred years have radically transformed the original situations resulting thereby in an increase in efficiency heretofore inconceivable.

The role of the spontaneously grown biosphere, subject to the organic rhythms of nature, has been preempted on a universal scale by a technosphere functioning according to inorganic and mechanical principles. This technological transformation of the material world has a double function; it frees man from traditional limitations and creates new possibilities for action. At the same time, however, the concrete physical existence of the technological artifacts and the functional laws of technological processes contain their own limitations. In place of the struggle with a hostile nature to be mastered, there is now a dependency upon a self-created technological environment which is largely taken for granted, but without which mankind could hardly survive. Thus we are not only forced to adapt to the structure of technological processes if we want to profit from them, but we are largely dependent upon the fact *that* such technological products are actually available.

b. *The Changed Situation of Life*

The shaping force of the technological environment has led to a general change of individual and social life. This phenomenon, too, is not fundamentally new, for man, as a natural species, has always been dependent on the material circumstances of life determined by the state of development of technology. Through industrialization, however, a decisive change has occurred — technology has become the *dominant* variable to which all other factors are subordinated. In previous times, too, the conduct of life and the possibilities of concrete action were limited by the prevailing technological conditions. Within these limits, however, individual and social life, human ideals, and the self-image of each era were defined by means of *non-technological* determinants. Thus a very summary characterization of classical times would include as key concepts ethical excellency and service for the political community. For the Middle Ages, faith and salvation of the soul were the essential points of reference. In contrast, our times are characterized by overt activity and a striving for quantitative and material progress, which have led, because of the continuously increasing action potential of technology, to a complete transformation of both the external and internal conditions of life (cf. Arendt *4.1*, 322). Here technological progress is considered as the highest value and as the absolutely binding point of reference, while the social, political, ethical, or esthetic consequences of technologization are considered as the results of an apparently unalterable and naturally given process that must be accepted as a matter of course.

The consequences are everywhere obvious. They extend from the levelling of consumer habits and tastes by means of mass production and standardization via planning and rational organization of the personal and social spheres all the way to the growing power of the technocratic state. The increased complexity of technological systems and procedures necessitates increased specialization and division of labor, a process in which rationalization and automation make possible a reduction of work time and simultaneously lead to a situation in which physical work becomes increasingly subordinated to organizational and control functions. Through modern means of transportation (railroad, automobile, airplane) and the electronic media (radio, television), daily routines and the range of experience — at least as far as it is indirectly accessible — have fundamentally changed. Improved medical care, based to a large extent on technological instruments, has produced a population explosion in the developing countries and a considerably increased life expectancy in the industrialized nations.

In all areas of knowledge, and, in particular, the empirical sciences, the use of refined methods of investigation and the employment of technology has led to a continuous gain in factual knowledge and in understanding cause/effect relations, the documentation of which provides resources for practical utilization. Artistic contents and forms of expression are also shaped by modern technology: the mass media, pocket books, low cost reproductions of works of art and music, provide the opportunity for general access to cultural goods, but imply, at the same time, the danger of conforming to the taste of the broad public. Life in a technological world becomes the subject of artistic expression, with sensitive artists, in particular, portraying the threatening features of the present situation, while others use technological modular parts as a functional component of artistic technique itself. Even man's self-image is increasingly influenced by the model of technological processes: a visit to the physician is compared to the 'maintenance job' or 'inspection' of a car, and that individual unable to cope with the objective, impersonal, technology-oriented work-style, is said no longer to be 'functioning', only to be 'replaced' as a faulty part. Even in the realm of religion, the influence of technology makes itself felt: the modern world view, which is informed by science and technology, conflicts with traditional, mythically and symbolically expressed ideas, and the traditional views are adapted to technology-oriented concepts.

The dominant role of modern technology becomes particularly obvious in the senseless arms race between the superpowers, in which the most recent scientific and technological insights are immediately applied to new armament systems. This example shows, in pointed form, the problem of the present situation: through large-scale and systematic research and development, in which one can relie on the entire fund of existing technological knowledge and capacities, ever more sophisticated technological possibilities are generated and then realized at a constantly increasing pace. What is unusual in this process is not the mere fact that the present level of technology is fully utilized for military purposes — that has always been the case. The essential innovation consists in the conscious and purposive expansion of technological knowledge. It is only through technological progress created in this way that an acute threat to mankind's continued existence arises, in that each new potentiality is then used without hesitation and without consideration of ensuing consequences. It would therefore be an error to condemn without qualification technology as such, which had heretofore served military purposes. What is crucial is the magnitude of the effects of modern technology, which is quite beyond comparison with those of all other preceding eras in human history. It is for this reason that the present situation can be mastered only by a fundamental reorientation for which necessarily the previous framework of norms and values, established to meet quite different conditions, must be called into question.

Between the various manifestations of technology, there exists an inner logical coherence such that the technological life-style as a whole forms an indissolvable unity. Gehlen speaks in this context of a "super framework" consisting of three interdependent components: scientific research, technological application, and industrial utilization. In this case, both research and practical utilization of results are completely rationalized so as to serve production and consumption with optimal efficiency, the result being that the entire technological process develops autonomously to a large extent (*5.8*, 12, 69—70).

In contrast to this interpretation, which focuses on concrete *action performances*, Freyer supplies an analysis of the *social structure* of a technological world. He observes that the present social structure constitutes a derived "secondary system". This system did not historically evolve, but was rather constructed exclusively for the purposes of meeting the requirements of smooth (technological) functions. This impersonal approach of universal planning and projecting includes social engineering: without regard for personal individuality and interpersonal relationships, man is completely reduced to his capacity as worker or consumer within the omnipresent, anonymous social mechanism (5.5, 79–93).

These two complementary positions express essential characteristics of modern technology: the increment in theoretical knowledge and its practical application takes place in a quasi-autonomous process which has as its prerequisite among other things, the accumulation structure of technological knowledge and ability and the multiplier effect of all technological innovations. To be able to fully exhaust any technological potential, man must adapt himself to the intrinsic requirements of technological action, so that he is considered, within the framework of a universal technology which dominates all areas of life, in the end only as functional element of technological processes.

The interrelationship of technological subsystems, in the last analysis, is conditioned by the historical decision to take the road of systematic utilization of the forces of nature and of science-based technological development. Implicitly following from this decision are all the functional conditions necessary in order to obtain the desired technological result. Since this approach has been maintained during the continued development of modern technology, the particular details of present-day technology fit into the framework outlined by it. Therefore, the aspect mentioned, which concerns mainly the concrete, physical or material side of technology, needs complementing by means of an appropriate human, decision-making component. As a second factor, then, we must add the basic willingness to utilize the created technological potential in every instance to obtain maximum yield. While it is true that this attitude can, in its root, be reduced to the basic need for the alleviation of the material conditions of life, nevertheless, the tenacity with which the chosen road is followed, even in the face of adverse results, indicates that a basic belief in material and technological progress is actually the underlying driving force. By way of brief summary, the reasons for the emergence of a uniform, self-contained technology can be found in the scientific and rational appropriation of nature and in beliefs about quantitative progress.

Such a discursive analysis, which gets its bearings from the techniques of engineering, i.e., from the analysis of structural and functional principles of concrete and material technological artifacts, contrasts sharply with Jacques Ellul's theory of the fundamental unity of all technological procedures — a unity which cannot be further analyzed conceptually. He works with a *broadly conceived* definition of technology which comprises all kinds of systematic and efficiency-oriented optimizations of means, distinguishing the categories of engineering techniques, economic techniques (labor science, planning), organizational techniques in government, politics, and the military, as well as human techniques (medicine, pedagogy, public media) (*5.4*, 22). In his opinion, all of these technologies can be adequately evaluated only as one coherent phenomenon, since any separate emphasis on one specific element would disregard the *intrinsic interconnectedness* of technology in its broadest definition (94).

As the English translator (of Ellul) emphasizes (*5.4, XIII*), Ellul is able, on account of this concept, to write a broadly designed phenomenology of technology which is comparable, in its claim to totality, to Hegel's *Phenomenology of Mind*. While he does not share with Hegel his method of historical genesis — Ellul's analysis is primarily aimed at systematics — he does share the latter's teleological conception. For Ellul, all forms of technology, i.e., all methodically employed procedures, are immanently necessary elements of an emerging comprehensive domination by technology.

The fact that such a description is possible shows the overwhelming importance of technology for the present time. However, since no analytical differentiation is made, it remains basically unclear as to why this situation could arise. This may be attributed in particular to the fact that Ellul underestimates the dominant role of the material systems and processes made available through *engineering technology*, for it is precisely these concrete artifacts and law-governed processes upon which all other methodological and technological procedures depend.

c. *Worldwide Expansion*

Besides the radical transformation of the physical world and the basic change of lifestyles, the third universal characteristic of modern technology consists in its global expansion. While it is true that technological inventions have always been transferred from one culture to another, such exchanges took place very gradually and in the framework of reciprocity. In contrast, today's modern technology is spreading from America and Europe across the entire globe, at an accelerated pace and unidirectionally. As a consequence, the emergence of a world civilization is taking shape which is – at least outwardly – rather unified, levelling historical traditions and cultural differences.

In this process, the perfection of transportation and communication provided by modern technology plays a key role. Today, every spot on the earth may be reached in a short time; the close political and economic ties, globe-spanning mass tourism and instantaneous transmission of news have established such intimate contact that each important event is acknowledged world-wide without delay. The place of foreign politics is taken by 'internal world politics' in which virtually the entire world participates.

However, man's ability to assimilate and adequately assess this information is greatly overtaxed, since events occurring in foreign social and cultural contexts, though widely publicized by technology, and only superficially understood. Since the possibility for firsthand verification is lacking, a vast potential exists for the manipulation of information which even unrestricted reportage cannot completely eliminate. Once again the fact is underscored that in the modern technological world secondhand knowledge predominates: the place of directly understandable and personally experienced events, characteristic of agrarian societies, for example, or of the artisan stage of technology, is taken by secondary information, which can only offer a limited picture of the complex political, economic, and scientific-technological phenomena.

Regardless of the concrete conditions which formed in a long tradition, lifestyles in the industrial countries are almost necessarily imported along with modern technology. Leading the way is the 'American way of life'. Just as peasants, fishermen, and shepherds are, for example, everywhere similar by virtue of their specific vocations, so too, computer specialists, planning engineers, or television mechanics everywhere are likely to have the same intellectual profiles. As Boulding emphasizes, their world language is technical English, and their common ideology is a belief in scientific and technological progress. He sees the external signs of the technological "superculture" in airports, throughways, skyscrapers, television sets, refrigerators, hybrid corn and artificial fertilizers, birth control, and institutions of scientific and technological education. All of these elements are geographically interchangeable anywhere in the world (5.2, 347). In addition, technologization not only creates certain material structures, but also a very specific work style. The successful application of technological procedures requires an appropriate organizational bureaucracy, an attitude impersonal, unemotional, and performance-oriented, along with dependable and disciplined work behavior. In comparison with these general characteristics of technology, the specific cultural and intellectual background of a country, or its particular form of government, is frequently little more than an epiphenomenon.

This process of world-wide technologization is possible because technological apparatuses, as well as technological know-how, are based on real physical interrelationships which can be stated in objective and empirical terms. Their objective foundation lies exclusively in the structure of the material world, and, therefore, they can without difficulties be separated from their particular social-cultural origins and transferred to another setting. Nevertheless, technological processes occur only to the extent that they are appropriately initiated and controlled by man. Therefore, together with the actual *transfer of technology*, there always has to take place a certain *transfer of culture* which imparts the technology-oriented attitude, itself a necessity if the procedures in question are to be implemented as planned. Since this attitude developed in Europe, albeit gradually and in the course of a lengthy process, the European culture constitutes a natural soil for it. This is not necessarily the case with the developing countries. Thus, for the developing countries the question arises as to how they may participate in the desired benefits of modern technology without betraying, in the process, their own cultural and intellectual traditions, and thus losing their own identity. The kinds of difficulties which arise in this context can clearly be seen if one considers that the accelerated pace of technological change has led, in the case of the industrialized nations, to various forms of alienation. In the third world, where centuries of development are made up for in a few years, an increase in alienation is therefore probably unavoidable.

In the course of the history of mankind, there emerged various distinct advanced cultures, whose cultural and technological interchanges always remained limited to a specific geographic area. Compared with this, the global expansion of technological civilization constitutes a completely unique phenomenon in human

THE TECHNOLOGICAL WORLD 163

history. In this connection, Claude Lévi-Strauss states that different nations have emphasized different aspects in coping with their conditions of life. In principle, of course, the same problems have to be mastered everywhere: all human beings share some form or other of language, art, and religion, as well as social, political, and economic structures along with specific knowledge and technical skills. The differences between cultures can therefore be interpreted as options for certain values. Thus, for example, in the case of the Eskimos and Bedouins, particular stress is laid on the mastery of a hostile environment, while in India, the philosophical and religious conception of life is in the foreground, or, in the Islamic world, a systematic connection between all aspects of life is established (*5.26*, 26–29).

From this point of view the question arises as to which reasons are responsible for the world-wide spread of modern technology. If the theory of differing options applies, then, strictly speaking, it is to be expected that there are nations with only a limited interest in technologization who would therefore accept modern technology only partially. As Lévi-Strauss observes, European technology and civilization was first only *indirectly* imposed on the Third World in the course of colonization: in terms of military technology, European armaments proved clearly superior with the result that the subsequent military occupation brought with it trade settlements, plantations and missions, and thus European lifestyle. In the face of this imbalance of power these nations were themselves forced to acquire modern technology (31–32).

This, however, accounts for only part of the world-wide expansion of modern technology, since even where the need for 'self-defense' never existed, there are today aspirations for rapid acquisition of sophisticated technologies. This process – which is reinforced through the improved possibilities of transportation and communication, themselves also the result of modern technology – is certainly also caused in part by the pressing need for basic necessities. Thus, for example, the growing world population can only be fed through the application of improved agricultural technology. Furthermore, the desire to keep up militarily and economically with the developed countries probably plays a role, too. In both cases, however, the deeper reason lies in the superiority of modern technological methods which are higher in yield and achievement to those of more primitive methods. If one judges by the criterion of efficiency, then modern technology is naturally superior to its predecessors since it was explicitly developed with this criterion as its basis.

Nevertheless, it still remains an open question why all the nations of the world adopt the norm of technological progress well beyond mere considerations of basic needs. In terms of concrete results (mobility, information transfer, wide-spread luxury), modern technology provides services which potentially appeal to all mankind as creatures of nature and which therefore strike a resonant chord with human desires, driving up demand and producing in everyone a 'need' for the latest technological conveniences. Modern technology addresses one facet of human existence and can thus, potentially, find resonance everywhere. As soon as the path of intensive technologization is taken, demands outstrip technological capacity, thereby creating further need for technological goods and services. Together with the pressure to conform, emanating from a technologically superior environment – superior at least in terms of concrete possibilities for action – this increase in demand leads to a world-wide spread of technology. In this way all of mankind pointedly concentrates on one of its abilities, i.e., mastery of the physical environment. This process absorbs all its energies, with the consequence that other possibilities, cultural activity or contemplation, necessarily recede into the background. (Regarding the basic limits in all fields of human activity, cf. Bálint Balla *5.1*.)

7. THE BENEFITS OF TECHNOLOGY AND THEIR COST

Technology, like all other forms of individual or collective action, stands in a specific relation of tension: technological systems and procedures supply the means for accomplishing intended ends. In this process the concrete situation which is felt to be in need of change undergoes a transformation consistent with the intended goals. The change which is produced, however, does not always correspond to original intentions but can, in fact, create new intrinsic conditions and infringements. Since all technological processes are based on a restructuring of the material world, the contrast between intended goal and actual result − possibly not intended at all − is particularly obvious.

Thus, modern technology was, on the one hand, introduced as a means for making life easier. On the other hand, however, because of its concrete and objective character, and because it ultimately reflects the structure of the physical world, modern technology led to many new limitations on human freedom of action. Because man as a biological species always needs to master his physical environment, and because such mastery requires conformity to technological processes themselves, now the inner logic of the technological world replaces the previous limitations. True, the drudgery of hard physical labor is largely eliminated in the process. The price, for it, however, consists in adaptation to the principles of technological action along with the alienating effects which they unavoidably produce.

Just as it was not any one human which created technology as such, these anthropological effects, which go far beyond the satisfaction of physical need, were not one person's doing. The destruction of the natural environment and the depletion of raw materials and energy, certainly not intended, and at first practically disregarded, entered the general consciousness only after they had proceeded beyond a certain point. The same is true for other secondary effects. Additional factors besides the actual effects are probably also relevant such as demand levels or awareness of the problems: in the developing countries, for example, or in post-war reconstruction such as that after World War II all energies are concentrated on establishing a basic level of material well-being with attention to negative side-effects necessarily sacrificed.

The unintended effects of modern technology are, however, so obvious today that different criteria of evaluation produce different emphases, without actually changing things. The occurrence of these effects may actually at first seem somewhat surprising since modern technology consists in general in the conscious and purposeful transformation and thus the 'humanization' of a hostile or at least humanly neutral nature (Norbert A. Luyten *1.2*, 141). In any case, the environment of technology ought generally to reflect human wishes and intentions. Furthermore, one could go on to argue that everything done in the framework of technology is of human creation and therefore could not be alien to mankind, or to put the argument even stronger: following the *verum factum* principle, which states that "we know only what we create", actually nature, that has been assimilated through technological systems and processes should be especially close and completely transparent to man.

A world that has been transformed in accordance with human needs should really make possible a high degree of self-realization and autonomy. In actuality, however, there are widespread complaints about technology these days — pertaining to alienation and to the heteronomy of the 'intrinsic exigencies of technology'. Such judgments are based on an unrelieved tension between the horizons of expectation and what is actually realized. Consequently, in every critique of technology it is possible to analytically abstract the following two components: (1) the description of normative ideals and, (2) the depiction of the actual state of affairs.

(1) Through explicitly stated or tacitly implied *ideals*, a certain

state of affairs, obtainable by means of technology, is distinguished as being desirable. Up to the present time, expectations about technology have tended to reflect an optimistic belief in progress and an unlimited faith in the possibilities of perfection of the human condition. The roots of this attitude can be traced, in the history of ideas, to the secularization of the Christian expectation of salvation. While in the medieval, religious view of the world, a consciousness of the basic inadequacy of all earthly activities was decisive, the modern shift of perspective away from ideas of transcendence led to a reinterpretation of the hereafter as being a historically attainable future for mankind on this earth. Heavenly paradise thus became earthly utopia. And what had originally been a mythological and eschatological idea was rationally reformulated into the Enlightenment idea of progress, according to which technology was to serve as an external aid for the improvement of human conditions. This view has been reinforced by the indisputable successes of technology such that, in the end, a faith in progress could spread to the extent that it is still exerting its influence well in excess of any reasonable justification.

This unlimited trust in the power of technology is first of all related to the possibilities of appropriation of the *physical* world by man. In very recent times, however, the concrete consequences of some extreme cases of technological development have shown that intrusions into the ecosystem and exploitation of supplies of raw material and energy cannot continue without serious repercussions. Also, with respect to *human* relationships, the limits of technological and organizational feasibility are becoming increasingly obvious: in the realm of the individual, essential relationships between persons (marriage, love, friendship, loyalty, comradeship) are based on an immediate bond which involves the person in its totality and which by its very nature cannot be consciously intended or systematically planned (Freyer 5.5, 84). And material affluence, in particular, which was made feasible by technology, has illustrated that the solution to all external problems leaves man largely untouched in his inner being, i.e., in his affections, aversions, responsibilities, and his guilt. Thus Fromm is correct in observing, "The assumption that the problems, conflicts, and tragedies between man and man will disappear if there are no materially unfulfilled needs is a childish daydream" (5.7, 106—107). An analogous situation is found in social problems where even the most perfect institutions, organizational structures, and technological aids can do no more than provide an external framework, the actual animation of which requires real people with their own personal opinions, wishes, and interests.

Assessments of the effects of technology, degrees of disappointment, or levels of satisfaction, reflect expectation levels themselves. Thus, while one's norms inevitably exceed what is actually the case, a large portion of current widespread criticism of technology reflects excessively high expectations in its ability to perfect the human condition. The attitude of technological optimism, which was justified in the construction phase of technologization in the industrialized nations, now appears inappropriate in the face of many instances of technological oversaturation. Nevertheless, it would be a mistake to fall toward the other extreme of technological pessimism. Only a sober investigation of the limits and possibilities of technology will yield a balanced assessment.

(2) The starting point for such an unprejudiced analysis is an investigation of *real systems and processes* on which technological actions are based. It must be borne in mind, however, that technology, like all other historical creations, always confronts man in a particular concrete form. It is thus one of the forms of life (language, ethics, science, religion, social institutions being others) into which each new generation grows, which it modifies, and then transmits. This constitutes the framework within which all human existence can assume its concrete shape. Just as the linguistic, cultural, and intellectual views supply one's self-image, values,

and ideals on which to base life, technology provides the objects and procedures for appropriating the physical environment.

Accordingly, a certain amount of historically mediated, technology-caused objectification inescapably accompanies the human condition: it is only through the externalization of his subjectivity by means of concrete processes that man can create the prerequisites for the material assurance of life and for all those higher cultural achievements which are based on it. Man is thus at once both creator and creature of technology. The manner in which each era utilizes the inherited technological milieu and further develops it is always an expression of its own self-image and its creative energy. By means of technology and the concrete world created by it, the living of one's personal life is necessarily affected, so that one unavoidably becomes dependent on the technology which one has oneself created.

The alienation of the industrial worker has been clearly pointed out by Marx: the worker objectifies himself in the production process and in the object of his work without developing a personal relationship to his activity and to the object produced (*4.17*b, 63–193). In a broader sense, life in industrial society — a life divided into many facets, therefore disjointed and subject to an extreme number of demands — leads to a situation in which people perceive their own existence as something foreign, not actually belonging to them, and with which they are unable to identify. One must, however, subtract a fundamental and existential alienation which is constitutive of every life situation, for man is always confronted with the task of struggling with certain given conditions, even in the best of cases.

The actual problem is thus reduced to the specific issue of *technology-related* alienation, which is characteristic of the modern technological world. This type of alienation, however, is based on the nature of technological procedures themselves, and can therefore be eliminated only to the extent that one is willing to do without the benefits of modern technology, too (Wolfgang Kluxen *5.20*, 84, 85). One can certainly avoid extreme cases, such as totally monotonous work on an assembly line, by means of different technological procedures. These usually entail added expenses, however. In principle, nevertheless, the price of the services of modern technology seems to include a degree of alienation. Against the immanent logic of unchanging technological processes, man must maintain the interplay of exertion and relaxation, which alone is appropriate for him. The same holds for all other technological issues: on the one side, there is the need for understandable and meaningful relations and processes. On the other, one is faced with the principles of specialization and division of labor, upon which the complex systems and processes of modern technology are based. All striving for authentic individuality is in contrast to standardization and interchangeability of parts and the uniformity they produce while making possible rational production and utilization of technological objects.

Thus, in the last analysis, technology-related alienation turns out to be a self-chosen fate which could only be altered by reducing the degree of technologization itself. Whoever is in principle opposed to rigidity, depersonalization, division of labor, and consideration of technological norms must logically also reject the services of technology. There appears to be little inclination to do this, however. Among even the most resolute critics of technology, hardly anyone is willing to voluntarily forgo the amenities and conveniences offered by modern technology. The common disparity between theoretical stance and actual practice must in fact be viewed as a vote in favor of technology. Yet, it would be a fundamental error to resort to some type of 'conspiracy theory', thus blaming certain groups, particular systems of political domination, or economic organization for what is actually due to the intrinsic logic of technology itself. Since these phenomena occur whenever technological action potential is efficiently utilized, alienation is bound to occur in *all* industrialized countries.

If one starts from the premise that technologization cannot be redone, the task at hand becomes the *coping with* rather the *abolition of* alienation. Seen in broad historical perspective, mankind has from its beginnings — for example, the transition from nomadic hunter to tiller of the soil — undergone all kinds of 'technology-related' transformations, in each case gradually internalizing the change until it became a new 'form of life'. From this point of view, it is the *acceleration* in present technological change that constitutes the actual problem, with alterations in the external conditions of life occurring so rapidly that neither the individual nor society is able to keep pace. Nevertheless, there is always the hope that "a human foundation may grow under technology" (Freyer *5.5*, 245) so that, beyond the mere factual interaction with technology, meaningful existence becomes possible. As an indication of this, for example, consider the matter-of-fact attitude with which new generations accept the technology-shaped environment and make use of its possibilities without surrendering completely to it. Just as is the case with physical problems of technology (environmental, material, and energy-related), the existential and cultural problems connected with technologization also require that the intensity and tempo of *future* technological development be channelled along acceptable lines (cf. Sachsse *5.33*, 70–74).

8. CHANGED CRITERIA

According to Johann Millendorfer, three phases of industrialization can be distinguished so far: in the first Industrial Revolution, the emphasis was on *material* problems which could be mastered with the help of new energy sources and new production methods. In the course of the ensuing development, an *informational* bottleneck developed to be subsequently overcome through the use of computers and new methods of organization and research. This constituted the second Industrial Revolution. Today a kind of third Industrial Revolution has become necessary, in which an answer needs to be found to the *ethical* question of how one is to live a meaningful life in a world shaped by technology (*5.30*, 408—13).

Millendorfer's schema makes it clear that different eras emphasize certain problems without thereby completely eliminating others of less importance. In actuality, for all phases of industrialization, certain material, informational, and ethical problems are *simultaneously* important, even though their objective significance and/or the subjectively felt importance carry different weights. Thus, at the present time, one can observe a growing awareness of ethical problems raised by technology. While one could start, in earlier stages of industrialization, from the tacit assumption that the technological action potential available must lead, in any event, to positive results and therefore must be fully utilized, today — at least in theoretical discussions — a certain skepticism and distancing from an unreflective optimism towards technology can be noticed.

Because all technological actions reflect certain *value concepts*, a rethinking and alteration in attitude is necessary, if any change in existing practice is to occur. In principle, such a change of course is always possible. While the situation as given does not permit any random change whatever, a range of options is available, and indeed necessary if new technological goals are to be sought. The putative inherent logic of things turns out to be no more than the expression of hypothetical imperatives which state how to proceed *if* certain values and ideals are to be realized (cf. above, Chapter III, Sec. 5).

The actual problem consists thus in using existing or future technological action possibilities in such a way that the well-being of mankind is actually served. Due to its generality and necessarily abstract and indeterminate character, this postulate will command general consent. The difficulties and disagreements arise as soon as one tries to derive specific instructions for action from it. One might point out, in this connection, that, from a historical perspective, the concrete conduct of life and the needs of the individual and society alike have always depended on the existing state of technological development. In that case, it is technology that forms man, and not *vice versa*. Since this fact applies as much to present times as to any other, one might further argue, an adaptation of the concrete performance of life and of the human self-image to technology is quite natural and should be welcomed.

In this connection, however, a distinction must be made. From a broad historical perspective, it indeed becomes obvious that the norms for a fulfilled life do not constitute unchanging entities that are independent of the milieu and of cultural and social tradition. This *global* observation about historical − and thereby also technological − conditioning of the self-image of each era in no way changes the fact that in each *concrete* historical context, the ideals of being human are to a large extent fixed. Thus, in our time in particular, the values of dignity, freedom, and personal integrity − based on cultural tradition rather than technology − are always declared if not always realized in practice. One must further remember that all technological acts intended to be more than ends in themselves always serve particular goals which, in principle, must be established independently from intrinsic technological issues. While it is true that the existing state of technology determines the range of potential action, it is not the outer-directedness by technological *facts* but rather the self-determination by fundamental *norms* that is ultimately decisive for human action.

Nevertheless, the discrepancy between the ever-expanding potential of technology and the largely static ethical concepts is quite obvious. This becomes especially apparent in the case of armament technology where the deadly arsenal of weapons, which has been brought to an advanced level with all means of scientific and technological ingenuity, has by now assumed apocalyptic dimensions. In the employment of this destructive potential, exceeding as it does all traditional norms, the law of the stronger applies as well today as it did for the Stone Age. The situation is similar in all other areas of technology: once natural forces were being systematically and scientifically exploited, a road was taken that has led, through the mechanism of accumulation and self-reinforcement, to steadily increasing magnitudes of technological effects. Since the consequences of present-day technology are graver and more far-reaching than in previous times, any irresponsible, egoistic, or special-interest action weighs much more heavily. Admittedly, individual and social decisions have always reflected short-range and selfish criteria. Apparently people have never exhibited that measure of circumspection and responsibility that ideally one might expect; thus, for example, available technological possibilities have always so far been used for military purposes or for the stabilization of political power systems. The dilemma, however, consists in the fact that, due to the increased technological effects, any present action taken according to traditional standards will produce utterly catastrophic results.

In view of this situation, in the limiting case an equilibrium between technological potential and ethical standards is conceivable in one of two ways: one could either reduce technology to a level commensurate with existing ethical norms or conversely develop ethical conceptions that suitably reflect the problems at hand. For either approach, a process of metatheoretical reflection about norms is mandatory, according to which contemplated actions are regarded from the outset as possessing ethical import. In both cases, a rethinking of the norms of action is called for: since modern technology is based on a certain attitude toward the physical world, toward the rationalization of procedures, and toward material progress, it can be changed only through a change in the pertinent values.

Herein lies the real problem. It is comparatively easy to theoretically outline new values which could safeguard the utilization of technology for human well-being. The difficulty lies in the fact that hypothetically generated ideals are neither obligatory nor actually effective. On the other hand, established values that are practically relevant are simply accepted in an unreflective manner as a matter of course. The acceleration of technological progress and the urgency of the problems we face do not permit our customary reliance on the gradual change of values that occurred more or less unnoticed and spontaneously. Thus, even in the area of ethics, a realm far removed from the *external* effects of technology and concerned with the *inner* core of human existence, the indirect influence of technology is unmistakable. The problems thus raised force man to re-examine traditional values in terms of a systematically integrated canon of values which itself, in a broader sense, constitutes a technological-methodological approach.

This shows once more man can, figuratively speaking, fight modern technology only with its own weapons. Just as unwanted physical effects of technologization can be alleviated only by appropriate technological counter-measures, so, too, in the ethical field, desired results can only be obtained through a systematic and rational approach. The dangers created in this way are obvious. Even if, following Ribeiro, one advocates, with however well-meaning and human intentions, "systematic intervention in the values that guide personal behavior" (*4.24*, 195), this is nevertheless the first step on the road to manipulation by technocrats. Taken to its logical conclusion, this direction from outside can lead to only one thing, a perfect dictatorship. On the other hand, one must acknowledge the fact that the intelligentsia has the responsibility in the current crisis to contribute such models to the present discussion as are capable of showing the direction in which a solution is to be found. Nevertheless, in a free society the acceptance or rejection of new values cannot be enforced through manipulative techniques or external force. The final decision can only rest with the persons that are themselves involved. It is they who must ultimately put the relevant norms into practice.

9. NEW VALUES

Any discussion about changed attitudes toward technology must begin with the accepted fact that the wheels of history cannot be reversed. Technology has become an integral part of our world, shaping both attitudes and daily life. Quite apart from the many forms of facilitation and convenience no one wants to give up, mankind's very existence is now dependent on appropriate technological measures. Because of these complex procedures and systems, our world has become so susceptible to breakdowns that an abrupt change would lead unavoidably to a general collapse. Consequently, taking the present as the point of departure, we must seek to channel technological development by means of gradual changes.

A starting point for this might be a renunciation of *technological* values in favor of *human* values. However, modern technological productivity requires an impersonal perspective, focusing totally on causal factors of physical processes. Any other approach, particularly an individualistic or 'subjective' one, would only interfere. To relinquish a style of thought which aims at anonymous and unemotional matter-of-factness, smooth functioning, and systematic predictability in favor of a concern for human values such as personal development and fulfillment or interpersonal concern cannot help but cause technological output to suffer.

A total withdrawal into contemplative introspection would not only be a flight from the world in which we live, but would also destroy the foundations of our material existence. Since we cannot do without technology, we must seek a balance between technology-oriented values, which are currently so prominent, and those which concern man as an individual. Discontentment with the present situation is reflected in numerous ways ranging from the emergence of the 'flower children' to the discussions about 'quality of life'. Although these different approaches vary widely and frequently contain obvious irrational elements, they all express a position of protest that, in due time, may lead to the required reorientation.

The one-sided fixation with 'progress' in terms of the *engineering sciences* was appropriate as long as one could hope that increases in technical efficiency would necessarily lead to *cultural* and *social* progress. It is precisely this hope that is no longer justified. Therefore, strictly speaking, every current innovation in technology ought to be assessed in terms of its social desirability. Conceivably, this could lead to the abandonment of some technological improvements because they were, on the whole, more bad than good, e.g., because they produced unacceptable environmental effects. Certainly there are numerous practical and theoretical difficulties in the evaluation procedures of "technology assessment" (Hetman *5.15*). For example, the *actual* effects of implementation of an envisaged technology cannot be predicted with certainty, nor is there agreement as to relevant *norms* or *institutional* and *organizational* decision procedures that should be employed. Nevertheless, given the far-reaching consequences of technologization today, an assessment and evaluation of technology that goes beyond short-term technical considerations has become unavoidable.

One may continue to advocate continuous positive *progressive development* even while subordinating technical progress to social and cultural criteria. By contrast, a more fundamental critique argues that the all-encompassing forces of accelerated technological change be countered by a program of *preservation of the existing state of affairs* and conservation of tradition. Because change always occurs in a field of tension between tradition and innovation, the crucial parameter is the extent and velocity of change itself. In this regard, the material preservation of the biosphere and the cultural preservation of a broad variety of forms of life, with their individual habitualized and internalized elements, are in competition with progress-oriented technological mind-sets ever pushing for the fastest and most radical changes possible and for global uniformity. It could be added that the need for relatively stable states, both physical and socio-cultural, follows already from purely functional considerations, since every change thought to be progressive is meaningful only to the extent that it is not so abrupt as to lead to individual and social alienation and that further the earth itself remains inhabitable.

Regardless of whether one takes as one's point of departure economic considerations and present trends in resource consumption (à la *Limits to Growth 5.28*), or whether one focuses, as Gruhl does (*5.11*), on the growing destruction of the biosphere, or whether one adopts Illich's approach of comparing concrete individual and social advantages of technology with the accompanying expenses (*5.17*), it has now become clear that technological action cannot continue indefinitely to be guided by considerations of quantitative and material progress alone. Although the specifics of each of these analyses may be questioned, their overall message is clear: in view of the far-reaching consequences of technologization, increases in technological feasibility must not automatically be equated with progress. *Self-limitation* is required, although this does not mean that things come to a complete standstill or that present technological achievements be forsaken. Such a limitation is not meant to bar all technological innovation, but rather to insure that decisions about technology await critical assessment of side-effects and long-range consequences.

While convincing in itself, this demand raises certain basic difficulties. The value criteria which heretofore determined human actions were limited to relatively separate and manageable social units. The context of action was thus directly understandable and

THE TECHNOLOGICAL WORLD 179

limited in its scope. A 'regionalized ethics' was quite appropriate for organic artisan technology, since the interference with natural processes was relatively minor. Modern technology, however, with its systematic and large-scale transformation of nature and its far-reaching effects can only be mastered by a 'global ethics' the norms of which reflect more general interrelationships of higher-order systems. Our traditional, firmly established values and attitudes that evolved gradually over time are hardly sufficient to cope with the current situation.

Ideally, a basic feeling of responsibility toward the whole of mankind, including future generations, should exist, as a check on the arms race, resource despoliation, and environmental pollution. Gehlen even thinks that man is morally overtaxed if he is expected to consider spatially and temporally distant matters and higher-order effects in the same way as the concretely perceivable effects of his actions (*5.9*, 136–137). This limitation on the historically given present suggests that all considerations of the 'ought' must begin with an analysis of the 'is'.

It must be conceded that every instance of wrong attitude leads unavoidably to corresponding physical consequences, given the material and objective character of technological systems and processes. As a result, such negative effects finally 'mandate' the appropriate utilization of technological action potential: if the physical environment becomes uninhabitable, or if raw materials and energy reserves are completely used up, some radically new solution will *have* to be found. Nevertheless, mere reactive behavior that is dictated by the course of events is no substitute at all for prudent anticipatory measures, considering the far-reaching, irreversible effects of technology.

As explained in Chapter IV, Section 8, modern technology can in principle be considered to be the result of an impersonal, rational, and worldly striving for success, the ascetic character of which Weber has noted (*4.29*). On the other hand, the many conveniences and luxuries afforded by technology are clearly antiascetic in character, striking a resonant chord world-wide and fostering the general expansion of technology itself. A constantly rising level of demand with expectations that the future will see even greater technological achievements is the very opposite of an ascetic attitude. It is thus clear that the asceticism and concentration of effort on mastery of nature which gave rise to modern technology needs to be recaptured: i.e., for technological development to be wholesome we need conscious abandonment and voluntary renunciation of what is merely feasible.

10. THE CRISIS IN THE ASSESSMENT OF TECHNOLOGY

One important reason for the wavering and uncertain attitude toward modern technology lies in its uniqueness in the *universal history* of mankind. There is therefore a lack of sources of comparison which could provide orientation or a criterion for evaluation. In the analysis of the road which finally led to the present situation, it has been shown that modern technology is based on manyfold intellectual, technical, and socio-economic prerequisites. This leads to the question as to whether these conditions emerged merely by historical accident or whether, instead, modern technology *had* to be developed. Judgments about modern technology are crucially dependent on the way this question is answered, for if technology owes its existence to a chance constellation of circumstances, one could in principle issue a negative assessment on what has occurred as being, as it were, an unfortunate accident of world history. However, as soon as one sees modern technology as a necessary stage in the history of mankind, a positive judgement is a necessary consequence, unless one is to reject the entire course of historical development — a move that, given the fact that one's own existence is itself grounded in a historical context, amounts to self-negation or to escapism.

Whenever questions are raised as to why technologization did not begin earlier, or 'which obstacles and inhibitions hindered technological progress' (Rüstow *4.25*, 61–62), it is tacitly presupposed that we are dealing with inexorable processes, which had to prevail in the end against any and all resistance. Such a manner of thinking is quite appropriate as a methodological device for highlighting certain factors or illustrating the sequence of different stages. What cannot be verified, however, is the ontological thesis that the development of technology is a predetermined teleological process which leads irrevocably to an increase of technological perfection and which, along the same lines, should actually have matured sooner. In view of the vastly different and unique preconditions which converged and led ultimately to industrial technology, we can only address the question of why modern technology occurred when it did, and not why it did not happen sooner.

In contrast, Riberiro sees in the technological innovations of European industrialization "natural and necessary phases of human progress which were *bound** to occur somewhere, e.g., if not in Europe, possibly in Islamic, Chinese, or Indian regions" (*4.24*, 199, cf. also Lévi-Strauss *5.26*, 38–39). In view of the fact that there were no significant changes in the most varied advanced cultures over very long time spans, however, this assertion appears extremely problematical. The opinion of Diesel is more modest, when he states: "If all culture were destroyed, then the entire development of technology as we have experienced it, from the beginnings of culture through present times, *could** recur, based on the condition of man on the earth and on his qualities" (*5.3*, 19). His thesis that the *potential* for technology is always in man is based in particular on the fact that even the most modern technological creations were predicted long ago.

As, for example, the transfer of technology to the Third World

---

* My italics (F. R.).

shows, it is indeed comparatively easy to *repeat* the process of technologization *if* it has already originated elsewhere. All technological processes are subject to laws of the physical world; their effectiveness is a reflection of objective and material conditions. Therefore, they can be repeated at will and transferred into different cultural contexts. Because of this fact, the possibility exists, in principle, for the world-wide expansion of modern technological procedures, with the superiority of the 'more advanced' technology being obvious for all to see.

From the present-day triumph of industrialization, it does not follow that those conditions which made it happen had to arise. Since each advancement speaks for itself, one can observe an overall progression in technology, a process enhanced by cross-cultural exchanges. If, however, one considers the fact that in different cultures technology remained practically unchanged for long periods of time, the possibility is at least conceivable that technology might remain at the artisan level, meeting immediate economic needs and never crossing the threshold into industrialization.

In the last analysis, however, such *hypothetical* speculation about historical necessity — that is, the question of the historical metaphysical legitimation of modern technology — must be left open. In terms of what *actually* happened these are moot questions anyway, since the systematic transformation of the material world by science-based technology is an accepted fact, and since, furthermore, there is every indication that we are only at the beginning of a universal 'technological age'. In this respect, world history, or in Hegelian terms, the world tribunal, has rendered its verdict: human development can no longer be imagined without technology; by all indications it is more than just a passing historical episode (Zimmerli *1.4*, 157). More specifically, the process of comprehensive technologization has created undreamt-of possibilities which promise to increase exponentially as long as it is pursued. At the same time, however, it becomes ever clearer that unreflective optimism toward progress is unjustified in the fact of the excesses of technology. While it is true that a groping awareness of limits is emerging, mankind is not yet ready to face the fact that conscious self-limitation is required or else technology's 'blessings' could become technology's 'curse'. Nor is there agreement about exactly where such limits lie. Thus, technology which was first perceived as positive and worthy of every effort, and was therefore systematically created, has begun to appear problematic. But it is unclear how future developments ought to be reasonably channelled.

In this general crisis of technological consciousness we are directed — since the world-historical perspective offers no reliable guidance — to the heritage of the European tradition which is, after all, responsible for the origins of modern technology. Yet here a new difficulty arises. Because industrial technology aims systematically and with scientific methods at increased technological efficiency, all traditional approaches and beliefs only stand in the way of progress. This holds just as much for technical limitations in engineering as it does for social, economic, or cultural situatons. Industrial technology, which arose in opposition to traditional procedures and economic forms, is intrinsically *hostile to tradition*. The historical context is always viewed as nothing more than an annoying limitation that should be dispensed with as fast as possible and never as a basis to be preserved and built upon.

Beyond this, world-wide technologization has more and more become a *supra-historical* phenomenon, determining the future in independence of the achievements of great persons or of cultural and national characteristics. Yet, it only *seems* as if the historical dependency of all human conditions could be eliminated by technology. Since the consciousness of one's own historical position is lacking, the suppressed historical relation asserts itself negatively and, as it were, behind the back: the present is devalued in favor of an idealized past or a future full of promise. The concrete situation of life, in which human existence occurs and in which alone it can find fulfillment, thus acquires its actual relative value only through reference to a state that is no longer or not yet present.

This escape, however, is no solution. Today's dominant values, it is true, are historically mediated just as are the factual conditions of life. They are to this extent necessarily shaped by the past. And, on the other hand, the far-reaching and momentous technological measures require prudent planning with future possibilities taken into account. Any romantic nostalgia for an ostensibly whole, untainted, and more humane world, however, turns out to be, upon closer inspection, a self-deception. That world was actually harsh, inhuman, and full of suffering; the idyll possesses its romantic halo only in idealized projection. We are dealing with the same type of thinking, only in reverse with respect to time, whenever the unsatisfactory present is contrasted with some utopian future society which has been perfected by technology and in which all contradictions are supposedly eliminated. Compared with such exaggerated expectations, the present appears as nothing more than a relatively insignificant transitional stage lacking the possibility for authentic and autonomous existence.

In either conception, the present is deprived of its true value and legitimate importance. Nevertheless, the present critical transition phase cannot be fully mastered without reference to the past or consideration of the future. What is truly disturbing and irritating about the present is the steadily accelerating pace of technological change that affects all areas of life — a pace which, although created intentionally and systematic in its details, and despite its origin in concrete human values and action processes, nevertheless seems, on the whole, to be willed by fate. As H. G. Wells has aptly expressed it in the final sentence of his *The War in the Air*: "Somebody somewhere ought to have stopped something, but who or how or why were all beyond his ken" (*5.38*, 395). Robert Musil goes further in stating that "mathematics, the mother of the exact natural sciences, the grandmother of engineering, was also the archmother of that spirit from which, in the end, poison gases and fighter aircraft have been born" (*5.32 I*, p. 41). And Lewis Mumford even speaks of the "Crime of Galileo" in banning personality and individuality from the scientific world view (*5.31*, 57–60).

A sober analysis shows, however, that no *single event* produced the specific intellectual prerequisites for modern technology. The intellectual foundation of what was concretely realized in the Industrial Revolution and in the ensuing process of technologization formed gradually in the course of extended development. The origins of this mode of thought are easily traceable as far back as classical antiquity. Thus, Max Bense correctly notes that the empirical approach of the late Scholastic period is just as relevant as are the artistic and scientific trends of the Renaissance or the accomplishments of Rationalism and the Enlightenment: "It can be concluded that the technological traditions are firmly rooted in the intellectual traditions of the West" (*1.7*, 221). In opposition to the radical critique quoted above, which sees technological thought as an aberration and a mistake, modern technology can, with good reasons, be regarded as a legitimate result of European intellectual and cultural history.

However, the rational style of thinking and the concept of perfected mastery of nature by no means exhausts the intellectual history. In Christianity and in Humanism, in Romanticism and in Existentialism, the unfolding and fulfillment of the individual existence is highlighted in one form or another. All of these currents of thought are genuine components of our traditions and any adequate critique of technology must reflect this fact.

The contradictory aspects of technology are poignantly evident in utopian literature: in the *positive utopia*, technology possesses healing power the use of which makes possible the appropriation of nature and the creation of the peaceable society. By contrast, *dystopian* writings describe the total manipulation of the personality, shaped by technology and functioning purely mechanically, void of intellect and soul. Modern technology thus turns out to be fundamentally ambivalent, both in origin and in future promise. The crisis in the present assessment of technology goes beyond superficial concerns with mere 'negative externalities' to a deeper awareness of this ambivalence. To grasp this means already to have taken the first step toward mastering the present situation.

BIBLIOGRAPHY

0. BIBLIOGRAPHICAL SOURCES

(a) General Works

0.1   Mitcham, C. and Mackey, R.: 'Bibliography of the Philosophy of Technology', in *Technology and Culture* **14** (April 1973); also as book: Chicago & London 1973. (Comprehensive, systematically ordered and annotated survey of literature; reviewed in *Technikgeschichte* **41** (1974), 353–355).
0.2   Herlitzius, E.: 'Technik und Philosophie', in *Informationsdienst Geschichte der Technik* (Dresden) **5** (1965), 1–36.
0.3   Sachsse, H. (ed.): *Technik und Gesellschaft*, 3 vols., Pullach & Munich 1974–1976. (In volume I, around 400 works on all aspects of technology are systematically ordered, analyzed, and discussed. The two remaining volumes provide annoted texts; volume 3 contains, among other things, an anthology on the philosophy of technology compiled by A. Huning).
0.4   Rammert, W.: *Technik, Technologie und technische Intelligenz in Geschichte und Gesellschaft*, Bielefeld 1975 (Report Wissenschaftsforschung). (Detailed commentary and literature survey on historical, sociological, and economic research on technology).

(b) Proceedings

0.5   *Akten des 14. Internationalen Kongresses für Philosophie – Wien 1968*, vol. 2, Vienna 1968.
0.6   *Natur und Geschichte – 10. Deutscher Kongreß für Philosophie – Kiel 1972*, Hamburg 1973.
0.7   *Proceedings of the 15th World Congress of Philosophy – Varna (Bulgaria) 1973*, vol. 1, Sofia 1973.

(c) Journals

0.8   *Technikgeschichte* (Düsseldorf).
0.9   *Technology and Culture* (Chicago).

BIBLIOGRAPHY

Detailed bibliographies are contained in *1.1*, *1.4*, *1.6*, *1.11*, *1.30* and *3.7*. Cf. further the survey on the development of the philosophy of technology in Chap. 1.

1. PHILOSOPHY OF TECHNOLOGY

(a) Readers

1.1 Lenk, H. and Moser, S. (eds.): *Techne-Technik-Technologie*, Pullach 1973. (Works on the present state of the discussion).
1.2 Luyten, N. A. (ed.): *Mensch und Technik*, Freiburg & Munich 1967. (Workshop on technology and theology).
1.3 Mitcham, C. and Mackey, R. (eds.): *Philosophy and Technology*. New York 1972. (Texts on all aspects of the philosophy of technology).
1.4 Zimmerli, W. C. (ed.): *Technik − Oder wissen wir, was wir tun?* Basle & Stuttgart 1976. (Various recent approaches).

(b) Monographs

1.5 Baruzzi, A.: *Mensch und Maschine − Das Denken sub specie machinae*, Munich 1973. (Investigation in the history of metaphysics).
1.6 Beck, H.: *Philosophie der Technik − Perspektiven zu Technik-Menschheit -Zukunft*, Trier 1969.(Based on [Christian] philosophical tradition).
1.7 Bense, M.: *Technische Existenz*, Stuttgart 1949. (Essays on the rationalistic legitimacy of modern technology).
1.8 Brinkmann, D.: *Mensch und Technik − Grundzüge einer Philosophie der Technik*, Bern 1946.
1.9 Cassirer, E.: 'Form und Technik', in *Kunst und Technik* (ed. by I. Kestenberg) Berlin 1930. (The place of technology within intellectual and cultural life).
1.10 Dessauer, F.: *Philosophie der Technik − Das Problem der Realisierung*, Bonn 1927. (The ontology of pre-existing solutions).
1.11 Dessauer, F.: *Der Streit um die Technik*, Frankfurt a.M. 1956. (Summary account with detailed bibliography; for both titles, cf. *1.30*).
1.12 Ducassé, P: *Les techniques et le philosophe*, Paris 1958. (The philosophical meaning of technological procedures).
1.13 Haverbeck, W. G.: *Das Ziel der Technik − Die Menschwerdung der Erde*, Olten 1965.
1.14 Heidegger, M.: *Basic Writings from 'Being and Time' (1927) to 'The Task of Thinking' (1964)*, New York 1977. (Contains 'The Question Concerning Technology', which is a translation of 'Die Frage nach der Technik' in *Die Technik und die Kehre*, Pfullingen 1962). (Interpretation of technology in terms of metaphysics of being).

BIBLIOGRAPHY

1.15 Hübner, K.: 'Einführung in die Diskussion philosophischer Aspekte der Technik', in *1.4*. (Purpose, method, and consequences of technology).
1.16 Huning, A.: *Das Schaffen des Ingenieurs – Beiträge zu einer Philosophie der Technik*, Düsseldorf 1974. (Introductory historical and systematic survey).
1.17 Jünger, E.: *Der Arbeiter – Herrschaft und Gestalt*, Hamburg 1932.
1.18 Jünger, F. G.: *The Failure of Technology: Perfection without Purpose*, Hinsdale, Ill. 1949. (Tr. of *Die Perfektion der Technik*, Frankfurt a.M. 1946).
1.19 Kapp, E.: *Grundlinien einer Philosophie der Technik – Zur Entstehungsgeschichte der Cultur aus neuen Gesichtspunkten*, Braunschweig 1877. (The earliest treatment).
1.20 Lenk, H. and Ropohl, G.: 'Praxisnahe Technikphilosophie', in *1.4* (Survey and outline of the historical development).
1.21 van Melsen, A.: *Science and Technology*, Pittsburgh 1961. (The nature of technology and its effect on man).
1.22 Meyer, H. J.: *Die Technisierung der Welt – Herkunft, Wesen und Gefahren*, Tübingen 1961.
1.23 Moser, H.: *Metaphysik einst und jetzt*, Berlin 1958; contains: 'Kritik der traditionellen Technikphilosophie'; reprinted in *1.1*. (Concerns especially *1.10* and *1.14*).
1.24 Moscovici, S.: *Essai sur l'histoire humaine de la nature*, Paris 1968. (An English translation under the title *The Human History of Nature* is forthcoming. [London, Harvester Press, and Atlantic Highlands, N. J., Humanities Press, in press]). (Technology as appropriation of nature).
1.25 .Ortega y Gasset, J.: *Betrachtungen über die Technik*, Stuttgart 1949. (The approach of cultural philosophy and anthropology).
1.26 Schröter, M.: 'Philosophie der Technik', in *Handbuch der Philosophie*, section 4, Munich and Berlin 1934 (reprint Munich & Vienna 1972). (Structural systematics and cultural significance of technology).
1.27 Simondon, G.: *Du mode de l'existence des objets techniques*, Paris 1969. (The genesis of technological objects and their cultural integration).
1.28 Spengler, O.: *Man and Technics – A Contribution to a Philosophy of Life*, New York, 1932. (Tr. of *Der Mensch und die Technik*, Munich 1931.)
1.29 Stork, H.: *Einführung in die Philosophie der Technik*, Darmstadt 1977. (Current manifestations and consequences of technology).
1.30 Tuchel, K.: *Die Philosophie der Technik bei Friedrich Dessauer – Ihre Entwicklung, Motive und Grenzen*, Frankfurt a.M. 1964. (Cf. *1.10* and *1.11*).

2. DIFFERING VERSIONS OF THE CONCEPT OF 'TECHNOLOGY'

2.1   Von Gottl-Ottlilienfeld, F.: *Wirtschaft und Technik* (Grundriß der Sozialökonomik, section 5), Tübingen 1914.
2.2   Seibicke, W.: *Technik – Versuch einer Geschichte der Wortfamilie τέχνη in Deutschland vom 16. Jahrhundert bis etwa 1830*, Düsseldorf 1968.
2.3   Skolimowski, H.: 'The Structure of Thinking in Technology', in *Technology and Culture* 7 (1966), 371–383; reprinted in *1.3* and *3.7*.
2.4   Teßmann, K.: 'Zur Bestimmung der Technik als gesellschaftlicher Erscheinung', in *Deutsche Zeitschrift für Philosophie* 15 (1967), 509–527. (On the philosophy of technology in the GDR [East Germany], see the special issues 1965 and 1973 of the same journal).
2.5   Tondl, L.: 'On the Concept of "Technology" and "Technological Sciences"', in *3.7*.
2.6   Tuchel, K.: 'Zum Verhältnis von Kybernetik, Wissenschaft und Technik', in *0.5*, vol. 2.

3. METHODOLOGICAL ANALYSIS

3.1   Agassi, J.: 'The Confusion between Science and Technology in the Standard Philosophies of Science', in *Technology and Culture* 7 (1966), 348–366; reprinted in *1.3* and *3.7*.
3.2   Du Bois-Reymond, A.: *Erfindung und Erfinder*, Berlin 1906.
3.3   Bunge, M.: *Scientific Research II – The Search for Truth*, Berlin & New York 1967.
3.4   Eyth, M.: *Lebendige Kräfte – Sieben Vorträge aus dem Gebiet der Technik*, Berlin 1905.
3.5   Hübner, K.: 'Zur Intentionalität der modernen Technik', in *Sprache im technischen Zeitalter* 25 (1968), 27–48.
3.6   Kotarbiński, T.: *Praxiology – An Introduction to the Science of Efficient Action*, Oxford & Warsaw 1965.
3.7   Rapp, F. (ed.): *Contributions to a Philosophy of Technology – Studies in the Structure of Thinking in the Technological Sciences*, Dordrecht & Boston 1974.
3.8   Ropohl, G.: 'Prolegomena zu einem neuen Entwurf der allgemeinen Technologie', in *1.1*.
3.9   Rumpf, H.: 'Wissenschaft und Technik', in *Die Philosophie und die Wissenschaften – S. Moser zum 65. Geburtstag* (ed. by E. Oldemeyer), Meisenheim 1967.
3.10 Von Wright, G. H.: *Handlung, Norm und Intention* (ed. by H. Poser), Berlin & New York 1977.

4. THE ROAD TO MODERN TECHNOLOGY

4.1 Arendt, H.: *The Human Condition*, Chicago 1958.
4.2 Bacon, F.: *Works* (ed. by J. Spedding et al.), London 1857–74.
4.3 Blumenberg, H.: *Säkularisierung und Selbstbehauptung* (New edition of parts 1 and 2 of 'Legitimität der Neuzeit'), Frankfurt a.M. 1974.
4.4 Brunner, O.: *Neue Wege der Verfassungs - und Sozialgeschichte*, Göttingen ²1968.
4.5 Burckhardt, J.: *Force and Freedom; Reflections on History*, New York 1955. (Tr. of *Weltgeschichtliche Betrachtungen*, Stuttgart 1969).
4.6 Daumas, M.: *A History of Technology and Invention: Progress through the Ages*, vol. 3, New York 1979. (Tr. of *Histoire générale des techniques*, vol. 3, Paris 1968).
4.7 Dijksterhuis, E. J.: *The Mechanization of the World Picture*, Oxford 1961. (Tr. of *De Mechanisering van het Wereldbeeld*, Amsterdam 1950).
4.8 Eliade, M.: *The Forge and the Crucible: The Origins and Structures of Alchemy*, 2nd ed. Chicago 1978. (Tr. of *Forgerons et Alchimistes*, Paris 1956).
4.9 Fleischer, H.: *Marx und Engels – Die philosophischen Grundlinien ihres Denkens*, Freiburg & Munich ² 1974.
4.10 Hall, A. R.: 'Engineering and the Scientific Revolution', in *Technology and Culture* 2 (1961), 333–341.
4.11 Hall, A. R.: *The Historical Relations of Science and Technology* (Inaugural Lectures, Imperial College of Science and Technology), London 1965.
4.12 a. Hoselitz, B. F.: *Sociological Aspects of Economic Growth*, Glencoe 1960.
b. Hoselitz, B. F.: 'Economic Development and Change in Social Values and Thought Patterns', in Zollschan, G. K. and W. Hirsch: *Social Change: Explorations, Diagnoses, and Conjectures*, Cambridge, 1976.
4.13 Hübner, K.: *Kritik der wissenschaftlichen Vernunft*, Freiburg & Munich 1978.
4.14 a. Kant, I.: *The Critique of Pure Reason* (tr. by Norman Kemp Smith), London 1929.
b. Paton, H. J.: *The Moral Law: Kant's Groundwork of the Metaphysic of Morals*, New York 1967 (used as source of quotation).
4.15 Klages, L.: *Der Geist als Widersacher der Seele*, 3 vols., Leipzig 1929–1932.
4.16 Klemm, F.: *A History of Western Technology*, Cambridge, Mass., 1964. (Tr. of *Technik – Eine Geschichte ihrer Probleme*, Freiburg & Munich 1954).
4.17 a. Marx, K.: *Capital* (tr. by E. Untermann), New York 1936.

b. Marx, K.: *The Economic and Philosophic Manuscripts of 1844* (ed., with an introd. by Dirk J. Struik), New York 1964.
4.18 Mathias, P. (ed.): *Science and Society 1600–1900*, Cambridge 1972.
4.19 Matschoß, C.: *Werner Siemens*, vol. 2, Berlin 1916.
4.20 Mennicken, P.: *Die Technik im Werden der Kultur*, Wolfenbüttel & Hannover 1947.
4.21 Mumford, L.: *Technics and Civilization*, New York 1934.
4.22 a. Nietzsche, F.: *'The Birth of Tragedy' and 'The Genealogy of Morals'* (tr. by Francis Golffing), Garden City, N.Y. 1956.
b. Nietzsche, F.: *Beyond Good and Evil* (tr., with commentary, by Walter Kaufmann), New York 1966.
4.23 Rapp, F.: 'Die Forschung in der Technik (bzw. Technologie) des 19. Jahrhunderts', in *Studien zur Wissenschaftstheorie des 19. Jahrhunderts*, vol. 12 (ed. by A. Diemer), Meisenheim 1978.
4.24 Ribeiro, D.: *Der zivilisatorische Prozeß*, Frankfurt a.M. 1971. (There is an English edition under the title *The Civilizational Process*, Washington, D. C. 1968. However, the German edition is based on the English, Spanish and Portuguese editions and contains important revisions of the earlier editions. References, therefore, are to the German edition).
4.25 Rüstow, A.: *Ortsbestimmung der Gegenwart*, vol. 3, Zürich & Stuttgart 1957.
4.26 Scheler, M.: *Problems of a Sociology of Knowledge*, London 1980. (This volume represents a translation of Chapter II – 'Die Soziologie des Wissens' of Scheler's *Die Wissensformen und die Gesellschaft* in *Gesammelte Werke*, vol. 8, Bern & Munich ²1960. References are to the English edition).
4.27 Sombart, W.: *Der moderne Kapitalismus*, vol. 1, Munich & Leipzig ³1919.
4.28 Stöcklein, A.: *Leitbilder der Technik – Biblische Tradition und technischer Fortschritt*, Munich 1969.
4.29 Weber, M.: *The Protestant Ethic and the Spirit of Capitalism*, 2nd ed., London, George Allen and Unwin, 1976. (Tr. of 'Die protestantische Ethik und der Geist des Kapitalismus', in *Gesammelte Aufsätze zur Religionssoziologie*, vol. 1, Tübingen, ²1922).
4.30 White, L., Jr.: *Medieval Technology and Social Change*. Oxford 1962.

5. THE TECHNOLOGICAL WORLD

5.1 Balla, B.: *Soziologie der Knappheit*, Stuttgart 1978.
5.2 Boulding, K. E.: 'The Interplay of Technology and Values', in *Values and the Future – The Impact of Technological Change on American Values* (ed. by K. Baier and N. Rescher), New York 1971.

BIBLIOGRAPHY

5.3   Diesel, E.: *Des Phänomen der Technik*, Leipzig & Berlin 1939.
5.4   Ellul, J.: *The Technological Society*, New York 1964.
5.5   Freyer, H.: *Theorie des gegenwärtigen Zeitalters*, Stuttgart 1955.
5.6   Freyer, M.: *Über das Dominantwerden technischer Kategorien in der Lebenswelt der industriellen Gesellschaft* (Abh. der Akad. der Wiss. und der Lit. in Mainz, Kl. 1960, Nr. 7), Wiesbaden 1960.
5.7   Fromm, E.: *The Revolution of Hope – Toward a Humanized Technology*, New York 1968.
5.8   Gehlen, A.: *Man in the Age of Technology*, New York 1980. (Tr. of *Die Seele im technischen Zeitalter – Sozialpsychologische Probleme in der industriellen Gesellschaft*, Hamburg 1957).
5.9   Gehlen, A.: *Anthropologische Forschung*, Reinbek 1961.
5.10 Glaser, W. R.: *Soziales und instrumentales Handeln – Probleme der Technologie bei A. Gehlen und J. Habermas*, Stuttgart 1972.
5.11 Gruhl, H.: *Ein Planet wird geplündert*, Frankfurt a.M. 1975.
5.12 Habermas, J.: *Technik und Wissenschaft als 'Ideologie'*, Frankfurt a.M. 1968. (Also contains the essay 'Erkenntnis und Interesse' available in English under the title *Knowledge and Human Interests*, Boston 1971).
5.13 Harich, W.: *Kommunismus ohne Wachstum?*, Reinbek 1975.
5.14 Heisenberg, W.: *The Physicist's Conception of Nature*, London 1958. (Tr. of *Das Naturbild der heutigen Physik*, Hamburg 1955).
5.15 Hetman, F.: *Society and the Assessment of Technology*, Paris 1973.
5.16 Horkheimer, H.: *Eclipse of Reason*, New York 1947.
5.17 Illich, I.: *Selbstbegrenzung – Eine politische Kritik der Technik*, Reinbek 1975. (This German edition was revised and enlarged by the author and is based on the French edition (Paris 1974), also enlarged by the author. The original English edition appeared under the title *Tools for Conviviality*, New York 1973. References are to the German edition.)
5.18 Jaspers, K.: *Man in the Modern Age*, Garden City, N.Y. 1957. (Tr. of *Die geistige Situation der Zeit*, Berlin & Leipzig 1931).
5.19 von Kleist, H.: 'Über das Marionettentheater', in *Sämtliche Werke und Briefe* (ed. by H. Sembdner), vol. 2, Munich 1965.
5.20 Kluxen, W.: 'Humane Existenz in einer technisch-wissenschaftlichen Welt', in *Technokratie und Bildung* (ed. by H. Janssen), Trier 1971.
5.21 Kranzberg, M. and Davenport, W. H. (eds.): *Technology and Culture*, New York 1972.
5.22 Krohn, W.: 'Zur soziologischen Interpretation der neuzeitlichen Wissenschaft', in Edgar Zilsel: *Die sozialen Ursprünge der neuzeitlichen Wissenschaft*, Frankfurt a.M. 1976.
5.23 Lem, S.: *Summa technologiae*, Frankfurt a.M. 1976.
5.24 Lenin, V. I.: *Collected Works*, vol. 31, Moscow, 1966.

5.25 Lenk, H. (ed.): *Technokratie als Ideologie*, Stuttgart 1973.
5.26 Lévi-Strauss, C.: *Race and History*, Paris 1952. (Tr. of *Race et Histoire*, Paris 1952).
5.27 Marcuse, H.: *One-Dimensional Man*, Boston 1964.
5.28 Meadows, D., et al.: *The Limits to Growth*, New York 1972.
5.29 Mesthene, E. G.: *Technological Change – Its Impact on Man and Society*, Cambridge 1970.
5.30 Millendorfer, J.: 'Konturen einer dritten industriellen Revolution', in *Stimmen der Zeit* **194** (1976), 408–419.
5.31 Mumford, L.: *The Pentagon Power (The Myth of the Machine*, vol. 2), New York 1970.
5.32 Musil, R.: *The Man without Qualities*, 2 vols., New York 1953–54. (Tr. of *Der Mann ohne Eigenschaften*, Hamburg 1952).
5.33 Sachsse, H.: *Technik und Verantwortung*, Freiburg 1972.
5.34 Schelsky, H.: *Der Mensch in der wissenschaftlichen Zivilisation*, Cologne & Opladen 1961.
5.35 Schumpeter, J. A.: *Capitalism, Socialism, and Democracy*, New York & London ^{2}1947.
5.36 Tondl, L.: 'Der Januskopf der Technik', in *0.5*.
5.37 Weingart, P.: *Wissenschaftssoziologie*, vol. *1*, Frankfurt a.M. 1972.
5.38 Wells, H. G.: *The War in the Air, and particularly how Mr. Bert Smallways fared while it lasted*, New York 1908.
5.39 Winterling, F.: 'Beziehungen zwischen Technik und Gesellschaft im utopischen Denken', in *0.3*, vol. 1.

NAME INDEX

Adorno, T. W. 14
Agassi, J. 58
Arendt, H. 12, 101, 103, 114, 120, 155
Aristotle 104

Bacon, F. 102, 107, 118, 148
Balla, B. 164
Baruzzi, A. 104
Beck, H. 33, 132, 133
Beckmann, J. 87, 92
Bense, M. 185
Blumenberg, H. 100
du Bois-Reymond, A. 5
Boulding, K. 110, 161
Brinkmann, D. 5
Brunner, O. 78, 100
Bunge, M. 58
Burckhardt, J. 30

Calvin, J. 95, 96
Carnot, S. 91, 92
Cassirer, E. 72, 73, 116, 122

Daumas, M. 85, 86
Davenport, W. H. 20
Descartes, R. 104
Dessauer, F. xi, 6, 10, 34
Diesel, E. 87, 121, 181
Dijksterhuis, E. J. 103, 104
Dilthey, W. 8
Ducassé, P. 66

Einstein, A. 43
Eliade, M. 71

Ellul, J. xii, 14, 71, 72, 120, 140, 159, 160
Eyth, M. 5, 34

Fleischer, H. 76
Francis of Assisi 74
Freyer, H. 13, 149, 157, 167, 171
Fromm, E. 168

Galileo, G. 83, 105, 115, 185
Gehlen, A. 11f, 26, 27, 111, 112, 113, 114, 157, 179
Glaser, W. R. 149
v. Goethe, J. W. 103, 114
v. Gottl-Ottlilienfeld, F. 34, 35, 63, 115
Gruhl, H. 17, 178

Habermas, J. 9, 14, 15, 16
Hall, R. A. 85, 99
Harich, W. 18, 19, 151
Haverbeck, G. 132, 133
Hegel, G. F. W. 9, 15, 100, 159, 182
Heidegger, M. xi, 11
Heisenberg, W. 43, 124
Heron of Alexandria 80
Hetman, F. 177
Horkheimer, M. 14, 15
Hoselitz, B. F. 77, 78, 97
Hübner, K. 37, 68, 73
v. Humboldt, A. 103
Huning, A. 4
Huxley, A. 148
Huygens, Ch. 85

NAME INDEX

Illich, I. 18, 178

Jaspers, K. 9
Jünger, E. 10
Jünger, F. G. 10

Kant, I. 107, 131
Kapp, E. 4, 5
Klages, L. 113
v. Kleist, H. 74
Klemm, F. 80
Kluxen, W. 170
Kotarbiński, T. 8, 35, 47, 49, 51
Kranzberg, M. 20
Krohn, W. 16

Lem, S. 120, 128, 129, 130, 131, 132, 133
Lenin, W. I. 76
Lenk, H. 4, 14, 33, 70, 148
Leonardo da Vinci 33, 82
Lévi-Strauss, C. 136, 181
Luyten, N. A. 166

Mackey, R. 21
Marcuse, H. xii, 15, 16
Marx, K. 9, 15, 76, 115, 169
Mathias, P. 84
Meadows, D. 16
van Melsen, A. 132, 133
Mennicken, P. 98
Mesthene, E. G. 146, 147
Meyer, H. J. 120
Millendorfer, J. 171, 172
Mitcham, C. 21
Moscovici, S. 119, 120, 125, 126, 127, 128, 132, 133, 140
Moser, H. 4
Mumford, L. 95, 185
Musil, R. 185

Newton, J. 105
Nietzsche, F. 8, 24, 67

Oresmus, N. 104
Ortega y Gasset, J. 10, 26, 71
Orwell, G. 148

Rapp, F. 8, 88, 91
Ribeiro, D. 12, 29, 30, 31, 120, 175, 181
Ropohl, G. 4, 33, 50, 70
Rothacker, E. 8
Rumpf, H. 26
Rüstow, A. 82, 98, 181

Sachsse, H. 21, 34, 171
Scheler, M. 8, 9, 27, 93
Schelsky, H. 14, 15
Schmidt, A. 14
Schröter, M. 115
Schumpeter, J. A. 77, 79
Seibicke, W. 31
Shannon, C. E. 43
v. Siemens, W. 77, 91
Skolimowski, H. 34
Sombart, W. 76, 77, 83, 84, 86, 97
Spengler, O. 8, 51
Spranger, E. 8
Stöcklein, A. 102

Teßmann, K. 36
Thomas Aquinas 129
Tondl, L. 28, 35, 43
Tuchel, K. 6, 35

Weber, M. 95, 96, 97, 179
Weingart, P. 16
Wells, H. G. 184
White, L., Jr. 80, 82, 104
Winterling, F. 149
Wittgenstein, L. 24
v. Wright, G. H. 59, 69

Zimmerli, W. C. 182

SUBJECT INDEX

(see also the TABLE OF CONTENTS)

accumulation 136
action, technological 55
action theory model 41
activity, goal oriented 32–34
adaption to technological change 120
age of the mill 86
alienation 2, 13, 15, 122, 144, 145, 165, 169
alternative technology 60
analysis, logical 7
animal rationale 3, 95
animistic view of nature 83
antiquity 80
apparatus 50
application of technological systems 52, 54
approach
    analytical 21
    scientific 88–92
armament systems 54, 156
arms race 156
art 156
artifacts 47, 54, 153
artisan technology 27, 37, 83f, 99, 120, 124
asceticism 95, 99, 179f
assessment of technology 20, 177, 180–186

balance of interests 150
bias, theoretical 2
biosphere 123, 138, 154
    preservation of 178

capitalist mode of production 76–80, 84
Christianity 95, 102
closed-loop process 12, 111
Club of Rome 16
colonisation 163
communism 19, 76, 100
complementarity of inventions 87
consistency, logical 43
control of technology 13, 147
criteria 63–65
Critical Theory 14
criticism of technology 166, 168
Cultures, Two 3
cybernetics 7

decision-theory model 38
deficiencies in human organs 12, 111–115
definition of technology 159
delegation of planning competency 150
democracy 79
developing countries 78, 96, 120, 166
development of the universe 125f
differences between cultures 163
distancing from natural processes 71f

efficiency 49, 147, 148
empiricism 85
engineering sciences 57f, 64, 87–92
ethical norms

197

change of 173–176
global and regionalized 179
entrepreneurs 76f
environment 17, 49, 70
evolution of technology 126–131
exact measurement 85
expert, role of the 150

forms of knowledge 9
freedom and control 146f
Frankfurt School 14

genetic manipulation 131
goals and means 60–62
  and values 141

historical data 109
history of philosophy 20
history of technology 68, 137
homo faber 3, 12, 117, 119
human choice, analysis of 143
hypothetical imperatives 57–62, 142

ideal and real factors 75–80, 93
ideology 13
imperatives, technological 14, 130
Industrial Revolution 1, 29, 56, 69, 84–87, 122
innovation 63
inorganic processes 104f
inspiration, intuitive 5
instruments 4, 50, 52
invention 5–7, 91f
  act of 4, 10, 63
  developmental 6
  mechanical 82f
  pioneering 6

knowledge, technological 44

laws of nature 43f, 67f

life plan 10
loss of individuality 9

Marxism, orthodox 141
machine, automated 28
machines 4, 28, 31, 50
managers and entrepreneurs 77
materialisation, phase of 5
meaning of technological actions 3
means, technological 12, 53
mechanization 9
metaphysical interpretation of
  technology xii, 110–117, 125–133
Middle Ages 81f, 100, 104

natural sciences 7, 53, 57f
nature and history 125f
nature, second 67–71, 124
needs 10, 62, 70, 109, 113

obstacles to technological progress 181
optimism, technological 4, 19
organ projection 5
organic nature 119
organic technology 104f, 121

perpetuum mobile 43, 111
philosophy of science 7
planning 144, 150
population explosion 2
preconditions, social and economic 75–80
preferences of the consumer 45f
prehistory 26
progress, idea of 167
progress, technological 13, 16, 49, 135

range of possibilities 42–47, 72
rationalism, economic 96f

rationality 100f
relief principle 49, 111–115, 145
religion 156
reorientation of values 152
research methodology 139
resources
    intellectual 47
    material 45
restrictions, political and legal 46

secularization 78, 100
selection mechanism 136
self-limitation 18, 108, 178
self-reinforcement of technology xiii, 136f
side effects 61
social engineering 158
social theory model 40
specialization 25, 88f, 123
spirit 133
    of the times 143
spiritualization of matter 132
state, change of 51
structural
    analysis 68
    description of technology 25
subject, acting 47
systematic character 137f
systems engineering 7

techno-evolution 128
technocracy 14, 15

technological
    experts 89
    processes 158
technosphere 123, 154
tools 4, 12, 26, 28, 31, 49
traditionalism 97
transfer
    of culture 162
    of technology 162, 182
transformation of states 50
trial and error 85

understanding
    theoretical 8
    scientific 44
utilization of natural forces 66
utopias, technological 149

value concepts 38–41
values
    human 176f
    technological 176f

will to technology 98
world civilization 160
world-historical perspective of technology 180–184
world-wide spread of technology 164
work 48, 94
    process 140

BOSTON STUDIES IN THE PHILOSOPHY OF SCIENCE

Editors:
ROBERT S. COHEN and MARX W. WARTOFSKY
(Boston University)

1. Marx W. Wartofsky (ed.), *Proceedings of the Boston Colloquium for the Philosophy of Science 1961-1962.* 1963.
2. Robert S. Cohen and Marx W. Wartofsky (eds.), *In Honor of Philipp Frank.* 1965.
3. Robert S. Cohen and Marx W. Wartofsky (eds.), *Proceedings of the Boston Colloquium for the Philosophy of Science 1964-1966. In Memory of Norwood Russell Hanson.* 1967.
4. Robert S. Cohen and Marx W. Wartofsky (eds.), *Proceedings of the Boston Colloquium for the Philosophy of Science 1966-1968.* 1969.
5. Robert S. Cohen and Marx W. Wartofsky (eds.), *Proceedings of the Boston Colloquium for the Philosophy of Science 1966-1968.* 1969.
6. Robert S. Cohen and Raymond J. Seeger (eds.), *Ernst Mach: Physicist and Philosopher.* 1970.
7. Milic Capek, *Bergson and Modern Physics.* 1971.
8. Roger C. Buck and Robert S. Cohen (eds.), *PSA 1970. In Memory of Rudolf Carnap.* 1971.
9. A. A. Zinov'ev, *Foundations of the Logical Theory of Scientific Knowledge (Complex Logic).* (Revised and enlarged English edition with an appendix by G. A. Smirnov, E. A. Sidorenka, A. M. Fedina, and L. A. Bobrova.) 1973.
10. Ladislav Tondl, *Scientific Procedures.* 1973.
11. R. J. Seeger and Robert S. Cohen (eds.), *Philosophical Foundations of Science.* 1974.
12. Adolf Grünbaum, *Philosophical Problems of Space and Time.* (Second, enlarged edition.) 1973.
13. Robert S. Cohen and Marx W. Wartofsky (eds.), *Logical and Epistemological Studies in Contemporary Physics.* 1973.
14. Robert S. Cohen and Marx W. Wartofsky (eds.), *Methodological and Historical Essays in the Natural and Social Sciences. Proceedings of the Boston Colloquium for the Philosophy of Science 1969-1972.* 1974.
15. Robert S. Cohen, J. J. Stachel and Marx W. Wartofsky (eds.), *For Dirk Struik. Scientific, Historical and Political Essays in Honor of Dirk Struik.* 1974.
16. Norman Geschwind, *Selected Papers on Language and the Brain.* 1974.
18. Peter Mittelstaedt, *Philosophical Problems of Modern Physics.* 1976.
19. Henry Mehlberg, *Time, Causality, and the Quantum Theory* (2 vols.). 1980.
20. Kenneth F. Schaffner and Robert S. Cohen (eds.), *Proceedings of the 1972 Biennial Meeting, Philosophy of Science Association.* 1974.
21. R. S. Cohen and J. J. Stachel (eds.), *Selected Papers of Léon Rosenfeld.* 1978.
22. Milic Capek (ed.), *The Concepts of Space and Time. Their Structure and Their Development.* 1976.
23. Marjorie Grene, *The Understanding of Nature. Essays in the Philosophy of Biology.* 1974.

24. Don Ihde, *Technics and Praxis. A Philosophy of Technology.* 1978.
25. Jaakko Hintikka and Unto Remes, *The Method of Analysis. Its Geometrical Origin and Its General Significance.* 1974.
26. John Emery Murdoch and Edith Dudley Sylla, *The Cultural Context of Medieval Learning.* 1975.
27. Marjorie Grene and Everett Mendelsohn (eds.), *Topics in the Philosophy of Biology.* 1976.
28. Joseph Agassi, *Science in Flux.* 1975.
29. Jerzy J. Wiatr (ed.), *Polish Essays in the Methodology of the Social Sciences.* 1979.
32. R. S. Cohen, C. A. Hooker, A. C. Michalos, and J. W. van Evra (eds.), *PSA 1974: Proceedings of the 1974 Biennial Meeting of the Philosophy of Science Association.* 1976.
33. Gerald Holton and William Blanpied (eds.), *Science and Its Public: The Changing Relationship.* 1976.
34. Mirko D. Grmek (ed.), *On Scientific Discovery.* 1980.
35. Stefan Amsterdamski, *Between Experience and Metaphysics. Philosophical Problems of the Evolution of Science.* 1975.
36. Mihailo Marković and Gajo Petrović (eds.), *Praxis. Yugoslav Essays in the Philosophy and Methodology of the Social Sciences.* 1979.
37. Hermann von Helmholtz: *Epistemological Writings. The Paul Hertz/Moritz Schlick Centenary Edition of 1921 with Notes and Commentary by the Editors.* (Newly translated by Malcolm F. Lowe. Edited, with an Introduction and Bibliography, by Robert S. Cohen and Yehuda Elkana.) 1977.
38. R. M. Martin, *Pragmatics, Truth, and Language.* 1979.
39. R. S. Cohen, P. K. Feyerabend, and M. W. Wartofsky (eds.), *Essays in Memory of Imre Lakatos.* 1976.
42. Humberto R. Maturana and Francisco J. Varela, *Autopoiesis and Cognition. The Realization of the Living.* 1980.
43. A. Kasher (ed.), *Language in Focus: Foundations, Methods and Systems. Essays Dedicated to Yehoshua Bar-Hillel.* 1976.
46. Peter L. Kapitza, *Experiment, Theory, Practice.* 1980.
47. Maria L. Dalla Chiara (ed.), *Italian Studies in the Philosophy of Science.* 1980.
48. Marx W. Wartofsky, *Models: Representation and the Scientific Understanding.* 1979.
50. Yehuda Fried and Joseph Agassi, *Paranoia: A Study in Diagnosis.* 1976.
51. Kurt H. Wolff, *Surrender and Catch: Experience and Inquiry Today.* 1976.
52. Karel Kosík, *Dialectics of the Concrete.* 1976.
53. Nelson Goodman, *The Structure of Appearance.* (Third edition.) 1977.
54. Herbert A. Simon, *Models of Discovery and Other Topics in the Methods of Science.* 1977.
55. Morris Lazerowitz, *The Language of Philosophy. Freud and Wittgenstein.* 1977.
56. Thomas Nickles (ed.), *Scientific Discovery, Logic, and Rationality.* 1980.
57. Joseph Margolis, *Persons and Minds. The Prospects of Nonreductive Materialism.* 1977.
58. Gerard Radnitzky and Gunnar Andersson (eds.), *Progress and Rationality in Science.* 1978.

59. Gerard Radnitzky and Gunnar Andersson (eds.), *The Structure and Development of Science*. 1979.
60. Thomas Nickles (ed.), *Scientific Discovery: Case Studies*. 1980.
61. Maurice A. Finocchiaro, *Galileo and the Art of Reasoning*. 1980.